CRISIS AVERTED

CRISIS AVERTED

THE HIDDEN SCIENCE OF
FIGHTING OUTBREAKS

CAITLIN RIVERS, PhD

VIKING

VIKING
An imprint of Penguin Random House LLC
penguinrandomhouse.com

LIBRARY OF CONGRESS CATALOGING-IN-PUBLICATION DATA
Names: Rivers, Caitlin, author.
Title: Crisis averted : the hidden science of fighting outbreaks / Caitlin Rivers.
Description: New York : Viking, [2024] | Includes bibliographical references and index.
Identifiers: LCCN 2024009354 (print) | LCCN 2024009355 (ebook) |
ISBN 9780593490792 (hardcover) | ISBN 9780593490808 (ebook) |
Subjects: MESH: Disease Outbreaks—prevention & control |
Disease Eradication | Epidemiologic Methods |
Public Health Practice | Social Determinants of Health
Classification: LCC RA643 (print) | LCC RA643 (ebook) | NLM WA 105 |
DDC 362.1—dc23/eng/20240524
LC record available at https://lccn.loc.gov/2024009354
LC ebook record available at https://lccn.loc.gov/2024009355

Printed in the United States of America
1st Printing

DESIGNED BY MEIGHAN CAVANAUGH

To the invisible public health heroes

who have shaped our world

CONTENTS

CRISIS AVERTED

WELL-BEING

There is a proverb I like: a healthy person has a thousand wishes, a sick person only one. Good health is a precious gift that is easy to take for granted. When it falters, recovery becomes a singular focus. At least, that was my perspective when I endured a difficult twin pregnancy fraught with serious complications.

In 2018, between my sixteenth and thirty-first weeks of pregnancy, I had approximately fifty fetal ultrasounds. It may have been more—at some point I lost track. In a box in my basement, I have piles and piles of ultrasound images, each showing two little babies tucked together, first spaciously and later snugly.

Identical twins are a miracle. Unlike dizygotic (fraternal) twins, monozygotic (identical) twinning is very rarely hereditary and not typically associated with fertility treatments.[1] Monozygotic pregnancies are completely spontaneous, and nobody fully knows why one zygote deigns to split into two.

Identical twin pregnancies can be extraordinarily dangerous for both the mother and the babies. Instead of each fetus receiving oxygen and nutrient-rich blood through individual placentas, as in fraternal twin pregnancies, my babies shared a single placenta. Sometimes this works well enough—but in our case, it did not. The share of blood and nutrients that sustains each fetus can become unequal, leaving one fetus with inadequate supply while the other becomes deluged, a condition known as twin-to-twin transfusion syndrome (TTTS).

My doctors detected our TTTS when I was barely into the second trimester. From then onward, we rode a long streak of misfortune. Without treatment, twin-to-twin transfusion syndrome is almost always fatal to both babies and can be dangerous for the mother—and even with treatment, the risks of fetal death, prematurity, or permanent disability are high.[2]

There is a surgical treatment, but it is not for the faint of heart. Then again, for those affected, there is really no choice. The treatment is this: In my case, a fetal surgeon (yes, such a job exists) injected fluid into my abdomen to create space. The surgeon then made a small incision near my belly button and inserted a fetoscope, or special camera, maneuvered the instrument past the fetuses and navigated to the shared placenta. She then used a second tool, a surgical laser, to coagulate the blood vessels on the surface of the placenta to separate the blood supply as much as possible. This is known as the Solomon technique, named for the story in the Bible of the king who adjudicates a dispute between two women, each laying claim to a baby.[3] Solomon, in a clever bid to reveal which woman is the rightful mother, announces that the baby will be split in two. He knows that the real mother would rather give up the baby than allow it to be killed. But in the case of twin-to-twin transfusion syndrome, splitting the blood supply giving both babies life is typically the only cure.

For most women who undergo the treatment, surgery resolves the TTTS. For us, it did not. It was another exceptionally unlucky roll of the dice. Following my surgery, I spent two weeks on bed rest, except my frequent medical appointments. On the fourteenth day, just when I was slated to put the terrifying episode behind me, the other shoe dropped. The babies developed another vanishingly rare complication. Not all the blood vessels on the placenta had been fully coagulated. Some invisible vessels, perhaps just below the surface, remained open. One baby was receiving an infusion of red blood cells "donated" from the other baby. The recipient became overloaded, while the donor grew acutely anemic from the blood loss. For this my anemic donor received two fetal blood transfusions and I underwent a second, urgent surgery. By the time my doctors completed the new round of treatments, my uterus had been perforated at least six times. The twin-to-twin transfusion syndrome was finally resolved, but it was not clear how long I would be able to maintain the pregnancy, given how many critical interventions I endured.

Illness has a way of warping time and space. For months, I commuted back and forth to the hospital several times a week, often daily, in a desperate bid to keep my twins growing. Nothing else existed for me during those months except the pregnancy. At home, I alternated between watching *The Great British Baking Show* first on the couch, then in bed, then perhaps back to the couch if I felt ambitious. At the hospital, I had a favorite parking spot, and I knew which alcove would have a wheelchair. The nurses knew my TV lineup and I knew their weekend plans. I had a half dozen specialists, and I counted thousands of ceiling tiles. Through it all, it was not clear that my babies would survive. My body was fragile, and my mind was preoccupied by fear, narrowing my world to a grinding, banal terror.

At twenty-four weeks, the edge of viability, I breathed a quiet sigh of relief. At twenty-eight weeks, when the odds of survival increase markedly, I grew hopeful. Maybe we would pull through. Then, at thirty-one weeks, I went into labor that could not be stopped. For any other pregnancy, that would have been distressingly early, but for us it was better than we had hoped. The odds of survival at that gestational age are very good. A win, but still a precarious one.

Delivery marked the end of a catastrophic pregnancy and the beginning of a traumatic NICU stay. But eventually, after more than two months in neonatal intensive care and approximately twenty middle-of-the night phone calls from the hospital, my husband and I took our babies home.

MY STORY IS UNIQUE, in that TTTS is exceedingly rare. The confrontation with illness and our own mortality, though, is anything but rare. The medical system is, after all, the backdrop of many of life's most poignant moments, from birth to death. Not only do we take our children for checkups and stop by the pharmacy after work, but many of us will face at least one difficult diagnosis that will bring us into too-close acquaintance with the dizzying world of doctors' appointments, diagnostic tests, and sleepless nights.

The medical system is also more than just our personal experiences, one patient or family at a time. It also consumes a significant chunk of the federal budget in the United States. According to the Congressional Budget Office, major health care programs totaled $1.5 trillion in mandatory spending in fiscal year 2021.[4] Medicare, the federal program that insures over 60 million seniors[5] and people with disabilities, had a net annual cost of around $700 billion[6] or around 10 percent of the federal budget in 2021.[7] This level of spending is similar to outlays on national defense and is second only to Social Security in terms of single-program spending.

But despite the fact that the medical system weighs heavily on our lives and budgets, it is not wholly—or even principally—responsible for securing and sustaining our good health. For the most part, medical providers tend to us only when something has gone wrong or is in danger of doing so. There is another half of the equation, which is all that keeps us feeling well. And despite what glossy magazines would have you believe, it's not the sprawling wellness industry. Instead, it's medicine's quieter cousin: public health. Public health is concerned with the health of the whole population. It is about creating the conditions for good health—the air we breathe, the food and water we consume, the diseases that spread between us. And good health is not, the World Health Organization notes, "merely the absence of disease or infirmity." Health is "a state of complete physical, mental and social well-being."[8]

Public health, by virtue of focusing on populations rather than individuals, trades in impersonal numbers. "There are ten new cases today," we'll say of an outbreak. Or: "the rate of infection has grown from 20 cases per 100,000 population to 35 per 100,000 population." In my role as an epidemiologist, I might note that identical twin pregnancies like mine occur in around 3 out of every 1,000 live births.[9] Of those, twin-to-twin transfusion syndrome develops in a minority, around 10 to 15 percent.[10] Those statistics communicate what is known as the "burden of disease," or the factors that determine the impacts of a disease on individuals and society.

But I am not just an epidemiologist. I am also a survivor, and from that perspective, I feel that those numbers obscure the fact that there are people behind those statistics, each of whom is undergoing their own unique and profoundly terrifying experience. For me, my pregnancy was beyond my worst nightmares. The fact that our triple survival is so rare is irrelevant to our suffering. If any of the three of us had died, the survivors would not be comforted by statistics. It is stories like mine that are behind the big numbers you will read in this book. When you encounter those statistics, I hope you will remember that they are composed entirely of real people, who each have hopes and dreams and fears.

I hope you will remember, too, that there are few visible markers of the accomplishments of public health. Instead, traces appear in the invisible fabric that envelops our modern lives—if you know where to look. We do not have the ready equivalent of highways to traverse or smartphones in almost every pocket to stand as monuments to the victories we have secured. Mostly, when I look around with my epidemiologist's eyes, it is absence that I notice. There is a common refrain in public health that captures this paradox: if we do our jobs right, nothing happens. An outbreak does not grow into an epidemic. A child does not go hungry. A would-be smoker never lights up. Public health creates and sustains miracles constantly—but because of the nature of our work, many successes go unnoticed. They are evident only in the absence of what suffering there might have otherwise been. I admire all that this absence represents. It is not often that such wondrous accomplishments are so close at hand. By making things *not* happen, public health changes the course of history.

For example, to me, the drinking water that flows from the tap and the waste effortlessly disposed of by toilets are marvels. I appreciate that standards and regulations that stretch from farm to table keep our food supply safe. We no longer choke on cigarette smoke in the subway, and workplace safety regulations have decreased the risk of being poisoned by chemicals while at our jobs. No longer is infancy considered especially perilous, and these days infant mortality can seem like a distant concept until it is your baby whose life is threatened. But the fact is, infant mortality used to be incredibly common.[11] And although it was expert medical care that kept my children and me alive, it was *public health* recommendations for universal prenatal care that ensured that I, and millions of other mothers, had access to the tests that detected complications. It was because of public health that I supplemented with folic acid, abstained from alcohol, and knew how much weight to gain while pregnant. It is because of public health that the rate of global infectious disease deaths (until the COVID-19 pandemic) has fallen steadily for three decades.[12] And we know all of this because of the work that public health does to track the data. Although economic, political, and sociocultural factors all play a role in shaping our well-being, it is often through public health programs that these forces touch our lives. At home and at school and at work, we live in a world shaped by public health.

On rare occasions, public health achieves something even more spectacular. Fearsome scourges are allowed to fade into memory. Smallpox, once a killer of millions, was wiped off the planet. Guinea worm, polio, and yaws are not far behind. Leprosy, scurvy, and diphtheria have all been beaten back. These afflictions once tormented humanity; now, they are preventable, treatable, and extinguished or nearly so. Through a combination of science and technology, boots-on-the-ground effort and sheer persistence, public health has already saved billions of humans from early death and allowed them to flourish in life.[13] It is these achievements that stop me in my tracks. They set the bar for what public health can accomplish, given sufficient vision, leadership, and resources.

And yet, there is still so much to do. While I marvel at how far we have come, I also know that we have miles to go before we sleep. Polio eradication has been ten yards from the finish line for over two decades now but hasn't made it over. Too many people in low-income and high-income countries alike go to bed hungry. Toxic chemicals abound in everything from drinking water to soil to food packaging. Although there are legitimate debates about how best to address those issues, too often voters and policymakers don't even advance to discussing *how* to solve them. Instead, we become stuck in arguments over whether we should even try.

The perennial question is: With what money? US president Joe Biden has a favorite saying: "Don't tell me what you value. Show me your budget—and I'll tell you what you value."[14]

For public health, the prognosis is not good. The annual expenditure for it is, at best, a rounding error in the national budget—or "budget dust," in the parlance of policy wonks. The year before the pandemic, the US Centers for Disease Control and Prevention (CDC) had a budget of roughly $7 billion in discretionary spending, or less than 1 percent of what is spent on the Centers for Medicare and Medicaid Services.[15] Much of the funding that Congress allocates to the CDC goes directly to state and local health departments to support public health services. As any public health officer can attest, that money does not go far when spread across every jurisdiction in the country. In fact, according to *KFF Health News*, over three in every four Americans live in a state that spends less than $100 per person per year on public health.[16] That sum goes to administering vaccines, tracking and controlling infectious diseases; inspecting restaurants, hotels, and daycares; responding to disasters; providing medical care to pregnant women; managing the opioid epidemic; educating people with chronic illness; and more. Despite the enormous spread of responsibilities, public health is exceeded by nearly every other budget line, from education ($15,037 per pupil in 2020) to parks and recreation ($155 per capita in state and local spending in 2020), fire protection ($175 per capita in state and local spending in 2020), and highways and roads ($620 per capita in state and local spending in 2020).[17]

WITH A BUDGET LIKE THAT, it's no wonder that most people have never actually met someone who works in public health. As an epidemiologist, I used to received a lot of puzzled looks when I introduced my profession. That stopped when the COVID-19 pandemic hit. The confusion was exchanged instead for minor incredulity or even pity, given how busy my colleagues and I became. I also began to get a lot of questions. Many sentences began with, "While I have you here . . ." Nearly everyone was eager to learn how to live and behave during the pandemic, and they jumped at the opportunity to get answers. Suddenly, an epidemic wasn't something medieval or foreign, it was here and now. Dr. Anthony Fauci was on the TV screen each night and jargon like "quarantine" threaded daily conversation. Several years on, it's clear that new familiarity is here to stay. As an epidemiologist, mine is a world unknown to most. But it has been met with greater curiosity since the pandemic, a silver lining on which I aim to build with the stories in this book.

MY CORNER OF PUBLIC HEALTH, the place from which I start in these pages, is a small one. When I began my master's degree, students were taught that the public health profession has five core components: biostatistics, environmental health, epidemiology (including infectious diseases), health policy and management, and social and behavioral sciences. Although I am an infectious disease epidemiologist and my work centers on epidemics and pandemics, I pay homage to the other disciplines throughout. There is more to say than I have captured here about the myriad contributions that other branches of public health have made to our lives, and if this book inspires you, I hope you will explore the others as well.

My goal is to make the threads of public health visible in the fabric of our lives. If more people could recognize them, I think there would be a greater appreciation for all that the field has brought us. But not all progress is forward. Even as we humans are solving problems with one hand, we are busy creating new ones with the other. Many of the pathogens with which we now contend are a consequence of our fractured relationship with our environment, as deforestation and wild animal trades lure viruses out of the forests and into cities.[18] The COVID-19 pandemic is a prominent example, but hardly the only one. Choices and behaviors can confound progress, too. Witness the recent anti-vax movement that has weakened protections against diseases once nearly conquered.[19] And new technologies like synthetic biology bring about both solutions to old problems and previously unforeseen vulnerabilities, should they be used for harm instead of good.[20] These challenges keep me up at night and guarantee that the work of making things not happen continues with as much urgency and energy as ever.

This book is about how we solve problems in public health, and the people who solve them. Each story has its own plot and hero, but together I hope that they tell the larger tale of public health's role in our daily life and reveal the inner workings of the quest for progress. It is also a collection of crises averted. We have narrowly dodged the worst in the past. By learning from those near-misses and considering the signals we need to look out for—including the threats we don't even yet know we should be tracking—we can build a healthier, more resilient future, together.

PERSEVERANCE

The headquarters of the World Health Organization (WHO) is tucked at the end of a long drive, just up the road from the United Nations in Geneva, Switzerland. It is an odd building with a facade like a shipping container balanced on a too-small block. It's also very small—much smaller than you might expect from an organization whose catchment area is the entire world. Just 2,400 people work at headquarters, a headcount about the size of my high school.[1] If you have ever seen pictures or press interviews held at WHO headquarters, nearly all are taken in a single location, in a tiny atrium next to a coffee shop.

Outside the building rests a cast bronze and stone statue of a man on one knee, holding a needle to the shoulder of a young girl.[2] Two individuals I imagine as the girl's parents appear to keep loving watch behind her. The statue commemorates one of humanity's greatest achievements: the eradication of smallpox. It was for centuries one of the most feared diseases in history. The virus was transmitted easily and killed quickly, sometimes wiping out entire families or communities.[3] These days, it's a distant memory.

Eradication is an epidemiological term that means the complete banishment of a pathogen. In 1980, smallpox was declared eradicated, and the horror and fright of smallpox has since been allowed to fade. That absence is a tremendous gift from a previous generation of public health professionals, and an achievement that dramatically demonstrates the profound potential of public health to change history's course.

It was no easy task. It took vision, tenacity, and perseverance to overcome enormous odds, including suspicion, prejudice, and ignorance. On the eve of the fiftieth anniversary of eradication, that is the enduring legacy of the smallpox story. Public health can do big things, even with a small headcount and modest budget. As you read of the feat, imagine what the field might do with real money.

ACROSS HISTORY, the virus known to science as *Variola* went by several aliases, all of which hint at its status as the most dreaded of plagues. In the 1500s, it was called the "great fire" by Mayans, and to others it was known as the "spotted death."[4] In England, it was called the "speckled monster."[5] William Osler, one of the founders of modern medicine, referred to it as a "terrible" disease, "one which fully justifies the horror and fright with which small-pox is associated in the public mind."[6] It did not matter if you were rich or poor, lived in the city or the country. Whatever its name, it invoked terror. In the twentieth century alone, well after the advent of an effective vaccine, the virus still killed an estimated 300 million people, a number just short of the current population of the United States.[7]

The most characteristic feature of smallpox was the distinctive rash, which usually appeared in the second week after exposure.[8] The lesions first emerged in the mouth and throat then spread to the face and then outward from there. Eventually, they covered the entire body. Victims also endured fever, pain, headaches, fatigue, and digestive symptoms.[9] In drawings and photos of sufferers, nearly every square inch of their skin is covered with raised, fluid-filled blisters, or "pox."[10] This form was known as ordinary smallpox, and it was the most common manifestation of the disease. Anywhere from 10 percent to 60 percent of victims of ordinary smallpox died, a death rate similar to untreated Ebola infection.[11]

There is a misconception that viruses capable of spreading easily must be mild—a misunderstanding that dogged the COVID-19 response, as new variants emerged in quick succession, some of which caused more severe illness than the original strain.[12] The epidemiology of smallpox proves that this assumption is patently, dangerously, false. The virus was spread through close contact, passing easily within families and between household members. It transmitted through droplets produced during coughing or sneezing, or through fluid that wept from open sores. Belongings like bedding and clothing could also carry infectious virus—a characteristic that was sometimes used to disguise bioweapons as gifts.[13]

So destructive was the virus that it was a geopolitical force, like what we might consider a threat to national security in modern times. Throughout the eighteenth century, waves of smallpox gripped cities with panicked terror. In colonial Boston, an epidemic in 1721 claimed the lives of nearly one in thirteen residents.[14] During a 1764 wave, schools closed and residents fled the city.[15] Red cloth flags were hung outside the homes of victims to warn others to stay away.[16] From the end of the seventeenth century and into the eighteenth century, smallpox killed Queen Mary II of England, as well as the monarchs of Austria, Spain, Russia, Sweden, and France, and the emperor of China.[17] The social and political turmoil resulting from each of these events, and the hundreds more like them that accrued over the ages, was immense.

Even for those who survived infection, many were permanently marked. Once healed, pox wounds often turned into deep, pitted scars. George Washington and Queen Elizabeth I were both known to have pockmarks.[18] There were other lasting effects, too. Although the historical record is not clear, scholars believe that as many as one in eleven survivors developed blindness[19] and that approximately one third of all cases of blindness in Europe were a result of smallpox infection.[20] The virus was also known to attack the bones, joints, and brain, leaving some survivors with severe and lifelong disabilities. Smallpox was, in short, a horror.

IT WENT ON LIKE THIS for most of history, until scientific advances tipped the balance in humanity's favor using an ingenious strategy: turning the virus against itself. Although the exact origin of variolation (the practice of deliberately causing mild infection to produce immunity) is lost to history, it was likely practiced in Africa, China, and India, perhaps even before the sixteenth century.[21] At first, the means were crude. Infectious material, like dried pox scabs, were harvested from people who had had a mild case of the disease. The scabs were then turned into a powder to be blown into the nose or scratched into the skin of someone in search of protection.[22] Ideally, the recipient would become ill, but not as severely as if they had left infection to chance. The immunity acquired from a mild bout protected against future encounters with the virus, including from strains that could prove deadly.

Not only was variolation icky, but it was also risky. Though less dangerous than randomly acquired infection, it was still a gamble. As many as one in fifty people who were deliberately inoculated died—astronomical compared to the safety of modern vaccines, but less than a tenth of what would be expected from natural infection.[23] What's worse, the risk extended beyond the variolated recipient. The virus used in the procedure could spread between people just like any other virus, meaning it could infect people who had not chosen inoculation.

This practice may sound startling to modern ears, but remember that chicken pox parties were common well into the 1990s.[24] Chicken pox causes mostly mild disease in children, but infection grows more dangerous as people age.[25] Adults are more than twenty times more likely to die from chicken pox infection than young children.[26] In the pre-vaccine era, some parents would arrange for their children to be purposefully exposed, in hopes that their children would "get it over with" and become immune before they grew to an age where they would be more vulnerable to severe illness. The practice mostly died out when a safe and effective vaccine was introduced in the US in 1995.[27] (Despite its name, chicken pox is not related to smallpox, though they both cause a rash illness.)

THE DESPERATE SEARCH FOR PROTECTION against smallpox became markedly less risky in the late 1700s thanks to the work of physician Edward Jenner. According to one oft-repeated version of the tale, Jenner noticed that milkmaids rarely developed smallpox and made a fantastically risky guess that cowpox pus was somehow to credit.[28] Instead of using scabs from humans with mild disease, Jenner began using pus (yes, pus) from the sores of cows infected with cowpox. The cowpox virus, we now know, is related to the smallpox virus, and infection from one protects against the other. It was this cross-protection that enabled milkmaids, who encountered cowpox in the course of their work, to avoid smallpox infection. Cowpox also had the advantage of being both much less deadly to humans and not capable of spreading from person to person.[29] Jenner's innovation quickly spread around the world, forming the basis of what would eventually become a successful eradication campaign.

Hair clippings from Blossom, the cow Jenner used to harvest infectious material, were on display at the Johns Hopkins Bloomberg School of Public Health, where I am faculty.[30] I paid a solemn visit to the exhibit to see it for myself. I saw wiry cow hair scattered about inside a small, black rimmed picture frame. It was just as inspiring as I expected, which is to say not especially. I could only laugh at the disappointing memento. It too neatly encapsulates the dilemma public health finds itself in: even the most profound successes are all but invisible, and it's left to storytellers to invoke the drama in a handful of hair.

ALMOST AS SOON AS there were vaccines, there were "anti-vaxxers." Then, as now, health officials understood that not only did vaccines protect the people who received them, they also provided indirect protection more broadly, by limiting the virus's spread.[31] This insight inspired requirements like the Vaccination Act of 1853, a United Kingdom law that required infants to be vaccinated against smallpox in their first months of life.[32] The vaccines were free to poor families, and parents who refused could be issued a fine or even imprisoned.

This, predictably, generated suspicion and backlash not unlike the responses to modern vaccine mandates. Some opponents protested the requirement as an infringement on personal liberty, while others focused on the risks and potential harms of the vaccine. One anti-vaccination group even produced a periodical titled *Vaccination Inquirer*—similar incarnations have circulated ever since.[33] The protests worked. By 1898, a new vaccination act was passed that allowed parents to exempt their children from the requirement.[34] This string of events exactly reflects the push and pull between modern vaccine requirements to protect both individuals and public health, while acknowledging skeptics' demand for autonomy. It is a balance that public health still struggles to navigate, and I find it both reassuring and disheartening that it's a tale as old as vaccines themselves.

FROM JENNER'S TIME UNTIL WELL into the twentieth century, these early vaccines tamped down smallpox transmission and saved countless lives—but there were serious gaps. The scattershot approach of vaccinating left many people unprotected, especially in remote or poor communities. There was no grand strategy for how vaccination should be used to solve the larger problem of smallpox transmission.

I see public health in a similar situation today. We have safe and effective vaccines for dozens of diseases like measles, rubella, pertussis, mumps, and HPV. The catch is that those vaccines are not available or accepted everywhere, whether because of refusal, cost, or remote location. Attempting to vaccinate randomly, at only the individual or even family level, is not going to make much progress against a disease that spreads rapidly and invisibly. There must be a grander plan to get shots in arms. A vaccine, as the saying goes, is not a vaccination.[35]

For smallpox, the first outlines of creating an ambitious plan emerged in 1953, when the first director-general of the World Health Organization, George Brock Chisholm, floated the idea of pursuing global eradication.[36] He broached the subject at the World Health Assembly (WHA), the annual meeting of the WHO member states. Chisholm was a persuasive, innovative leader. It was he who proposed the WHO's expansive definition of health as "not merely the absence of disease or infirmity," a grand vision that has animated the field ever since.[37] On the matter of smallpox eradication, though, Chisholm was ahead of his time. Health Assembly attendees rejected the proposal, considering it an impossible fantasy.[38] One delegate scolded that such an effort "might prove uneconomical and would not . . . add to the prestige of the Organization"—a prophesy startling in its error.[39]

I should note here that institutional prestige is not meant to be a top priority in public health. But in this case, it became a roadblock, and one reason that eradication was passed over in 1953. Moreover, delegates to the World Health Assembly from a number of countries, including the United States, the United Kingdom, India, and El Salvador, did not think smallpox warranted international coordination and should be dealt with on a regional or local level. Some delegates argued that the world was simply too vast, too populated. Eradication would mean reaching every single human, across every single country, in a momentous shared project. And the naysayers said it simply could never be done.

Chisholm did not seek reelection and finished his post as the first director-general later that year. His proposal to eradicate smallpox died a slow death, thanks to a mix of political timidity and inaction that will feel familiar to any watchers of politics before or since. Science is not enough; it must be matched with a will, and the funding, to bring it to the world.

MIRACULOUSLY, FIVE YEARS LATER at the eleventh convening of the WHA, the idea to pursue eradication was resurrected. The Soviet deputy minister of health and delegate to the WHA, Viktor Zhdanov, argued vigorously against protests that eradication was impossible. It was notable that Zhdanov was even in attendance. The world was in the throes of the Cold War at the time, and the USSR had only begun sending delegates to the WHA the year prior, after almost a decade of absence from active involvement.[40] Zhdanov did not waste the precious opportunity. He arrived at the meeting with a lengthy report that he had prepared himself, detailing why eradication was both feasible and pragmatic.[41] Shrewdly, his argument played to the pocketbooks of attendees as much as their benevolence.

The virus had already been eliminated in the Soviet Republic through organized mass vaccination, Zhdanov noted in his report. In other places, too, mass vaccination campaigns were bringing the virus to heel, with smallpox nearly gone from the Americas. But although local transmission was interrupted on Soviet shores, cases were regularly imported by travelers—a recognition that disease does not respect borders. As long as smallpox continued to circulate somewhere in the world, countries everywhere would need to continue to vaccinate their population and maintain costly surveillance systems. Why not, he argued, pursue elimination in every country in the world, thereby sparing everyone the trouble and expense of continuing to battle the disease? He supposed, with perhaps a touch of hubris, that the deed could be done in one short decade.[42]

. . .

As Zhdanov reasoned in his analysis, most diseases are not candidates for eradication. To be considered, there are four criteria that must be met.[43] First, a disease must cause obvious, identifiable symptoms so that even people with no medical training can recognize it. Smallpox victims suffered from a distinctive, blistery rash that was not easily confused with other diseases.

Second, a disease must not be capable of spreading when no symptoms are present. COVID-19 fails this test, as does influenza, because even people without symptoms can harbor the virus and pass it to others. In the case of smallpox, a person was only contagious after symptoms began and so it was clear when someone must be isolated to prevent transmission. Third, a disease must infect only humans. Elimination from the human population is a formidable enough challenge. Tamping out a virus from the wild animal population is even more intractable. Rabies, for example, will likely never be considered for eradication because it lurks in bats, as well as many other animal species. Smallpox, mercifully, infects only humans and thus passed the test. And perhaps the most important criterion for eradication is that public health officials must have some means to halt transmission. For easily transmitted viruses like the common cold that do not have a vaccine, not much can be done to keep them from spreading. For smallpox, the vaccine that began with pus from a cow made all the rest possible.

Through his tenacity, Zhdanov's bid in 1958 accomplished what Chisholm's did not. It was still not a full commitment to pursue eradication. Even Zhdanov could not muster that miracle. But the

Assembly did adopt a resolution for the World Health Organization director-general to study and report to the World Health Assembly Executive Board on the feasibility of an eradication campaign.[44] A similar resolution to the study of the issue had bogged down Chisholm's proposal—even today, "pigeonholing," or killing a piece of legislation by starving it of floor time, is a common way for proposed laws to die. But this time, the idea had support. By the twelfth World Health Assembly, delegates voted unanimously to adopt the resolution as a step toward eradication.[45]

The decision to take our collective fate into our own hands was one born of spectacular vision and ambition. In Zhdanov's time, the fields of microbiology and immunology were still in the early stages. Despite causing a massive pandemic in 1918, the influenza virus was not isolated in the laboratory until 1933.[46] Poliovirus was not cultured until 1949.[47] The transition from unhappy victims of infectious disease to active combatants was swift and decisive, unfolding over the course of just decades. By adopting a resolution to pursue eradication, the World Health Assembly and the WHO member states from across the world set their sights high—not just on fighting the war against smallpox, but on winning once and for all.

LIKE MANY WARS, the fight against smallpox turned on logistics. By the time the eradication campaign began in earnest, most rich countries had already squashed the virus through mass vaccination. The countries still wrestling with the virus were scattered throughout Asia and Africa, and they were among the poorest in the world. Few people in affected countries had access to routine medical or preventive care, and basic infrastructure like roads for vaccination teams to travel on, or census records to track who should be vaccinated, were patchy. To pursue the virus, eradicators had to give chase through an ingenious and relentless ground game, and they had to do so in all thirty-odd countries where the virus still spread for years on end.[48]

To make it happen, vaccination campaigns used a hyperlocal approach, but on a global scale. First, an advance team visited villages or neighborhoods with the goal of securing assent and support from the local leaders. The eradicators would explain to leaders, such as the chief or elder, that they wanted to protect residents from smallpox by vaccinating as many people as possible. Once support was secured and a date for the clinic was chosen, the advance team then moved onto the next area, going from one to the next to lay the foundation for the later vaccination tour.

Next, the vaccination team arrived on the agreed-upon date. By then, local leaders would have prepared for the visit by encouraging families to participate, including arranging for outlying families to travel to town. With that taken care of, the vaccination team was left to focus on reaching every man, woman, and child in the area in a single, well-coordinated visit. The teams themselves were tiny, commonly consisting of only two to eight people.[49] Yet, this group could vaccinate hundreds of people per hour.[50] Once everyone had been reached, the vaccination team would move to the next village by foot, motorbike, or truck. It was an extraordinarily labor-intensive process that stretched far into the densest jungles and most inhospitable deserts, an incredible feat of endurance and commitment that amazes and inspires me even today.

Technological advances helped, too, though not always in the ways you might expect. In 1967, operations became somewhat less onerous thanks to a simple invention that made the process of giving a vaccine faster and more reliable.

In the early days of eradication, smallpox vaccines were administered using a jet injector, an alarming-looking contraption that punctuated vaccine into the skin using a high-pressure burst of air instead of through a needle.[51] The contraptions were widely reviled by vaccinators and vaccinees alike. The injectors were tricky to use and prone to breaking down, forcing eradicators to spend time fussing with their equipment instead of vaccinating. For recipients, they were loud and somewhat painful. The devices were finally retired when an American scientist named Benjamin Rubin went back to basics. Rubin took a classic sewing needle and ground off the tip of the end that holds thread, leaving two open prongs exposed. Though simple, it proved to be a transformative innovation. The bifurcated needle, a tool still in use today, works by capturing a tiny suspension of vaccine between the prongs. When vaccinators use the needle, they first dip the needle into a vial of vaccine and then poke it into the recipient's skin numerous times.[52] The process is much faster and quieter compared to the jet injector, and it requires almost no training. It also uses less vaccine, which allows for more efficient use of the limited vaccine supply. As an added benefit, bifurcated needles are cheap and easy to sterilize for reuse. During the eradication campaign, a bundle of one thousand cost just five dollars.[53] The simplicity of Rubin's invention is credited with making it possible for the eradication campaign to vaccinate as many as two billion people.[54]

The last step in a vaccination campaign occurred several weeks after the first visits. A sweeper team retraced the steps of the previous teams, traveling from village to village and neighborhood to neighborhood. At each stop, they counted the number of residents who had a visible pox scar on their upper arm. This scar, known as a "take," would form at the site of a potent vaccination.[55] People who had never been vaccinated or had been vaccinated with a bad batch would be unmarked. The goal was for at least eight in ten residents to bear the scar.[56] That was considered the threshold for herd immunity, or the point at which too many people were immune for the virus to circulate readily.

IT IS TEMPTING TO IMAGINE eradication as a sprawling operation. The campaign did touch every corner of the globe, and the number of eradicators reached into the hundreds of thousands. But rather than operating as one giant machine, the campaign was structured as a network of independently managed programs, with each country overseeing and customizing the work in their area. In fact, almost 70 percent of the $315 million in funding that went to support the final years of the program came from the countries where smallpox still circulated, all of which were profoundly cash-strapped.[57] The fact that poor countries were willing to allocate funds to their eradication program, despite their limited resources, demonstrated a remarkable level of dedication and determination.

Overseeing it all was the Smallpox Eradication Unit headquarters in Geneva, Switzerland, right in the building where the statue paying homage to the program now stands.[58] The headquarters team was first established in 1967, and it was tiny by any standard. It was staffed by no more than nine people at any time, with the majority serving as fixers. Their role was to help with any issues that arose in the country or regional offices.[59] If the vaccine supply ran short, the headquarters staff found more. If a political leader refused to lend support, headquarters staff arranged for tea and a stern word. When the bifurcated needle was invented, headquarters staff ensured it got it into the hands of eradicators. If it needed doing, the team would find a way to get it done.

The program was led by Dr. Donald Ainslie Henderson, who went by his initials, D. A. He was an American who joined the post after serving as chief of viral disease surveillance at what is now known as the US Centers for Disease Control and Prevention.[60] Already a giant in public health when he took up the mantle of chief eradicator, Henderson went on to become one of the greatest heroes of the century. Although his name is not widely known outside the circles of public health, Henderson and Zhdanov and the thousands of hardworking people who served in the eradication program over the years are now credited with saving the most lives in history.

BY THE TIME THE HEADQUARTERS team was established, the extensive, laborious, and logistically complex approach of vaccinating as many as people as possible had carried the eradication campaign to elimination in dozens of countries. But in 1967 in thirty-three countries, mostly low-income countries in tropical regions where national governments struggled under colonialism or its recent dissolution, the virus continued to circulate.[61] An additional fourteen countries faced imported cases. The task of reaching every person on the planet was so onerous and complex that remote and under-served communities would see vaccination teams only infrequently. Year over year, new babies were born, growing the ranks of unvaccinated infants and children until there were enough vulnerable people for the virus to take hold. An estimated ten to fifteen million people were still infected each year, many of whom were never properly diagnosed.[62] The eradicators began to suspect that the strategy of mass vaccination, an approach that had carried the program through its first decade, would not propel it to the finish line. Unless vaccinators could cover even more ground to visit even the remotest communities regularly, maintaining very high levels of vaccine coverage proved all but impossible.

With the mass vaccination strategy sputtering, the eradicators had no choice but to devise a new strategy. Like many good ideas, the winning idea was both simple and deviously clever. Rather than doubling down on maintaining high vaccine coverage worldwide, epidemiologists pivoted to a strategy called "surveillance-containment."[63] The approach leveraged the insight that not everyone is at equal risk of becoming infected with smallpox, and so not everyone should be equally prioritized for vaccination. The people at very highest risk are contacts, or people who have been exposed to someone who is sick. If contacts could be identified very quickly and either vaccinated or placed into quarantine, perhaps chains of transmission could be broken in a more targeted way.

With this in mind, eradicators erected special disease-surveillance programs. They passed out flyers and hung posters encouraging community members and health workers to be on the lookout for the distinctive smallpox rash. Any suspected cases were to be reported to public health authorities for investigation and follow-up. As an added incentive, officials offered payment to people who reported cases.[64] Enlisting community members to be the eyes and ears of the eradication campaign supercharged disease surveillance efforts and allowed public health officials to find and act on as many cases of smallpox as possible.

When a potential smallpox case was reported, a team of epidemiologists and health workers quickly traveled to the affected village or neighborhood to investigate. If smallpox was confirmed, they immediately implemented a strategy called "ring vaccination."[65] This involved identifying, vaccinating, and quarantining every single person who had been exposed. Sometimes, the contacts of the contacts were vaccinated, too. The goal was to create a ring of immunity around everyone at risk to prevent the virus from spreading further. It was a demanding task. To work, the entire process of finding and vaccinating contacts had to be completed in a matter of days, a difficult task even in the best of circumstances. If an outbreak was too large to execute this strategy quickly, vaccinators would return to mass vaccination to reach as many people as possible.

The dedication and persistence of the eradicators during this period of the program is breathtaking. Dr. T. P. Jain, for example, was assigned to a surveillance-containment team in Assam, India. The area had experienced severe flooding, and "[i]nvestigation and containment of many of the outbreaks required wading from house to house in areas in which leeches were legion and snakes a problem."[66] In Ethiopia, a team of twelve Sudanese eradicators traveled for two months to reach an area that was wrought with civil unrest. According to one account,

> [T]he team had to carry with it all the petrol and most of the supplies needed. Sudanese pounds were acceptable currency for the purchase of food over half the distance; for the last part of the journey, the team members needed Ethiopian dollars, and to obtain them they sold a supply of blankets which they had brought with them just for this purpose. The "roads" over which they travelled had not been traversed for years. It was necessary for them to construct bridges and in many areas to walk ahead of the vehicles, clearing a path with large knives. In some places, the underbrush was so dense that it had to be burnt (on one occasion the flames nearly consumed one of the

vehicles). Mechanical breakdowns, poisonous snakes, wild animals and insects were daily problems. Nevertheless, they persisted in their journey, during which they contacted and vaccinated some 20,000 people but found no smallpox.[67]

I CAN ONLY IMAGINE that at this point, the temptation was surely for Henderson and his colleagues to call it "good enough." The eradicators had already made so much progress in reducing the burden of smallpox that the world had become a markedly different place compared to just two decades earlier. During the beginning of the eradication campaign, from the late 1950s to 1970, the number of countries reporting cases fell by over half.[68] By 1974, just a handful of countries reported cases, including Ethiopia, India, Pakistan, and Bangladesh.[69]

So often, as the footprint of a disease dwindles, so too does the commitment to continue the fight. But consider this: In infectious disease epidemiology, there are concentric circles of impact. The center circle represents benefit at the individual level, the domain of a single person. When someone is protected by vaccination, better health is already closer at hand than it was for most of human history. For the person at the center of that circle, the benefits are profound. At the population level, though, not much will have changed. One would-be link in the chain of transmission is broken, which is no small thing, but everyone else in the community is still vulnerable. The work of public health is to push those protective measures out to larger and larger circles of people, expanding access and impact as far and wide as possible. This is the principal difference between public health and medicine.

As more and more people become vaccinated, a virtuous cycle takes hold. So many chains of transmission are broken that the virus can hardly circulate. This widens the circle of protection even further, extending to people who are not directly protected because they did not, for example, get vaccinated. This is a concept known as herd immunity, a term borrowed from veterinary medicine.[70]

At this stage, if local transmission is halted altogether such that the virus is no longer present in the area, the disease is "eliminated."[71] Many diseases have been eliminated from high-income countries, some of which you may not have had occasion to think of because they are so rare. Rubella, diphtheria, and polio all used to circulate widely in high-income countries like the United States, but are now considered eliminated because cases occur only in people who became infected while traveling abroad. But elimination is not permanent. It can slip away at any point. Any erosion in the number of people who are immune gives the pathogen a toehold to make a comeback. I see this most clearly in the return of measles, which was nearly eliminated in the United States until the return of anti-vaccination sentiments made it possible for the virus to circulate domestically again.[72]

What's better than control or elimination, though ambitious in the extreme, is to "eradicate" the pathogen altogether, ending transmission everywhere on the globe.[73] This solves the problem forever, benefiting not just the current generation of people who witness the death of a disease, but also all future generations who are born into a safer, healthier world.

The financial returns from eradication compound as well. Every country, even those that can hardly afford it, must spend vast sums of money to maintain disease surveillance, vaccination, and other disease control measures when a pathogen is still circulating. After eradication, that funding can be redirected, put instead toward combating other diseases, creating another virtuous cycle. This was one of

Zhdanov's central points in his proposal to the World Health Assembly.[74] Eradication was not only possible, he argued, but cost-effective.

It is not just about the math. One of the highest ideals in public health is a propensity toward action, a fundamental commitment to preventing as many deaths and illnesses as possible.[75] It is one of the cornerstones of the field and was a source of inspiration for me when I chose to become an epidemiologist. We do not always live up to those ideals, as anyone who followed the dithering decision-making in the early days of the COVID-19 pandemic will attest. But for every stumble, there are a dozen examples of epidemiologists moving at the earliest inkling of danger, forgoing the easier option of waiting to see how bad things might get before deciding to act. Take, for example, one of the most famous stories of the lengths that people will go to in order to stop an outbreak, including stretching the limit of human endurance and beyond.

THE REMOTE OUTPOST of Nome, Alaska, was not an easy place to live in 1925. It was one of the most inhospitable environments on the planet, almost a thousand miles away from supplies during the long winter months and gripped by frigid temperatures most of the year. When winter descends, Nome is on its own. Which is why when children began dying a painful death, suffocated by a thick gray film that coated their throats, the situation was grave. The town doctor, Curtis Welch, suspected diphtheria, a bacterial disease that is now rare thanks to routine vaccination.[76] But in the early twentieth century, diphtheria was a dreaded killer of children. The bacteria spreads easily and up to one in ten victims die.[77] Even in those days, there was an antitoxin serum treatment that could rescue children from death—but the remote town's stores were limited.[78]

Nome officials sent an urgent dispatch requesting emergency assistance. State officials, recognizing the enormous stakes, broke into a flurry of discussion. A train could transport antitoxin to another northern town, Nenana, but that still put the supplies hundreds of miles from Nome and there were no passable roads between. A plane could cover the remaining distance, which would have the benefit of speed. Planes had only recently come into use in remote Alaska, and proponents were eager to prove them an asset in the emergency. But that plan, too, soon dissolved. The high winds and cold weather, with temperatures well below freezing, would make the flight impossible. That left one option: dogsled teams.[79]

Dog sledding is a traditional form of transport in Alaska, one that is still in use to this day. Generations of Alaska Natives have fixed sleds to long wooden skis that glide over top of the ice and snow. The sleds are pulled by teams of winter-hardy dogs with thick fluffy coats, with a driver or "musher" directing from the back of the sled. As late as the 1920s, dogsled teams were used to deliver mail across the sprawling state during winter months.[80] But the journey under consideration was something else entirely, a wildly ambitious and dangerous conceit. Teams would have to traverse 674 miles of wilderness in the dead of winter, with temperatures well below zero degrees Fahrenheit.[81] They would face mountains, blizzards, frozen tundra, and dangerous, icy rivers. And they would need to do it as quickly as possible, for all the while, the epidemic of diphtheria was killing the children of Nome.

Despite the extraordinary risks, there was never a question of whether to leave Nome to its fate. The same spirit that calls firefighters to run into burning buildings or soldiers to leave no man behind animates an essential and enduring commitment to halting outbreaks. Whether it is diphtheria, smallpox, or many of the outbreaks that threaten us today, inaction is not an option. And so, on January 27, 1925, the first team of ten sled dogs, led by musher "Wild Bill" Shannon, set out from Nenana, Alaska, once the serum arrived by train from Anchorage.[82] The team traveled overland for over fifty miles, driving hard for over fourteen hours to pass the precious package of diphtheria antitoxin to another dogsled team, who immediately set out on the next leg of the relay. And so it went, through punishing days and frigid nights. Twenty relay teams forded almost 700 miles across the largest and most northern state in the union, in the dead of winter, to deliver Nome's salvation.[83] It was a journey that would normally take over a week. The teams completed it in just five and a half days, despite facing a blizzard with hurricane-force winds.

Upon reaching the finish line at 5:30 in the morning with the precious package in tow, musher Gunnar Kaasen is said to have collapsed from exhaustion, but not before hugging Balto, his team lead dog, and offering him high praise: "damn fine dog."[84]

The journey came at a significant cost. Three mushers suffered frostbite, and at least four sled dogs perished during the journey, overcome by the grueling physical demands and extreme weather.[85] At least five children died before the serum arrived, and more deaths may have gone unrecorded among Alaska Native families.[86] But thanks to the courageous dogsled teams, the epidemic was stopped and hundreds of children in and around Nome, Alaska, were spared a painful death.

Some of the musher teams garnered fame for their feats. Kaasen, Balto, and the rest of the team went on a cross-country tour of the United States. Crowds of thousands flocked to pay homage to the heroes. One of the dogs lives on as a statue in New York City, where he remains a favorite of visiting schoolchildren. (When installed, the statue provoked protest by anti-vaxxers, further proof that resistance to vaccination has deep roots.[87]) Several movies and books have been made of the tale, including *A Long Way to Nome*[88] and *The Cruelest Miles*.[89] But perhaps the best-known legacy of the Nome serum run is the Iditarod Race, an annual dogsledding race that retraces the entire route from Anchorage to Nome.[90] These popular culture retellings make the Nome serum run one of the best-known adventures in the annals of public health. And that same spirit of service, commitment, and ingenuity that delivered diphtheria antitoxin to Nome also carried the smallpox eradication campaign to the ends of the Earth.

THE BLOOD, SWEAT, AND TEARS paid off. The last case of naturally acquired smallpox, a case of *Variola minor*, was recorded 1977, 18 years after the smallpox eradication program began its work and 181 years after Edward Jenner invented the modern smallpox vaccine.[91]

The last case in history, Ali Maow Maalin, worked as a cook in a Somalian hospital. He was exposed to the virus in the course of his work. A tribute to Maalin, who died in 2013, tells the story of how he became infected. According to the Global Polio Eradication Initiative:

> A man carrying two smallpox-infected children from a nomad encampment had been driving all day, looking for the local isolation camp. Taking wrong turn after another, he finally decided to stop and ask for directions. He did so at the hospital where Ali worked. "Ali didn't think about it twice—he jumped in the van and immediately offered to accompany the driver," Mahamud tells us. The driver then asked Ali if he had

been vaccinated, but Ali simply said: "Don't worry about that. Let's go."[92]

Ironically, although Maalin previously had worked as an eradicator, he had not been vaccinated because he feared the needle. Maalin recovered from his infection and went on to serve as a vaccinator in the polio eradication campaign.[93] "When I meet parents who refuse to give their children the polio vaccine, I tell them my story," Ali said in 2006. "I tell them how important these [polio] vaccines are. I tell them not to do something foolish like me."[94]

That should have been the final chapter in humanity's long saga with one of history's most prolific killers, until a shocking turn of events claimed its final victims.

ON AUGUST 24, 1978, Dr. Henry Bedson of the United Kingdom's University of Birmingham examined a sample of pus taken from an English woman with a serious rash. As one of the world's foremost pox virologists, Benson had seen the transition of smallpox from fearsome killer to extinct foe. Although eradication was a triumph that he shared with thousands of other scientists and public health professionals around the world, it was also bittersweet. Bedson's lab was due to close at the end of the year as part of the WHO's bid to drastically reduce the number of laboratories that handled the virus in the post-eradication era.

But even as his laboratory planned to shutter, Bedson made a startling and upsetting discovery. The specimen that he examined showed evidence of smallpox infection. The sample was taken from a woman who worked as a medical photographer at Bedson's own university. No cases of the disease had been diagnosed in England in years. Indeed, none had been found anywhere in the world since Maalin's infection a year prior. About two and a half weeks after her diagnosis, the patient, Janet Parker, died from her infection. Before her death at just forty years old, she passed the virus to her mother, who became ill but survived. The pair became the last-ever cases of smallpox on Earth—but how had Parker become ill?

Bedson knew.

Not long after he made the diagnosis, the eminent scientist took his own life.

Parker worked one floor above Bedson's laboratory. The room that she used to make telephone calls was linked by the building's air handling system to the laboratory where Bedson and his assistants handled the *Variola* virus. Laboratory staff were revaccinated every one to two years to ensure their immunity was strong. Parker was vaccinated in 1966, a dozen years prior to her infection. Immunity against smallpox is long-lasting, so Parker's breakthrough infection was something of an anomaly. But the bigger mystery was how she had been exposed to the virus in the first place.

The laboratory, an investigation later revealed, did not use the modern biosafety protocols recommended for handling dangerous pathogens. In fact, a team of three experts from the World Health Organization had visited the laboratory for an "informal" inspection earlier that year. The inspectors were troubled by the lack of control measures and urged Bedson to ensure the facility was either "upgraded to meet the Standard or discontinue work with [V]ariola at the earliest possible date."[95]

For Bedson, neither were feasible options. There was no sense in investing in costly upgrades when his laboratory was closing for good in just a few months. Nor could he countenance sacrificing the last chance to finish his life's work. In fact, his workload had accelerated as the deadline drew closer. What few biosafety measures were in place wore even thinner as his team raced to complete their research before the end of the year.

Bedson did not work without oversight. In addition to the World Health Organization's inspection, he was also subject to review by both his own university and the United Kingdom's Dangerous Pathogens Advisory Group (DPAG), both of which should have ensured that the practices in his lab and others like it complied with safety protocols. Although an inspection by DPAG found that Bedson's laboratory did not meet the requirements, the inspectors recommended that work there be allowed to continue. Several discretionary exemptions were extended to allow the laboratory to operate.[96]

It seems likely that the safeguards failed because of what experts in national security now understand can be the weak link in any chain of security controls: human foibles. Bedson was an insider, a well-respected and senior scientist who had previously participated in both oversight committees that were meant to check his work. These positions afforded him a status that was beyond reproach. The mistakes may have cost Janet Parker her life.

Today, only two laboratories in the world handle *Variola* virus, one at the Centers for Disease Control and Prevention in the United States and one in Russia.[97] But there are hundreds of laboratories around the world that handle other dangerous pathogens.[98] Biosafety and biosecurity policies and practices have advanced since Bedson's day, but they still rely heavily upon self-governance by supervising scientists and their peers. Considering the enormous number and variety of active laboratories and the relatively few associated incidents, this system works remarkably well. But when serious accidents or mishaps do occur, the consequences can be devastating.

· · ·

IT STILL ASTOUNDS ME THAT smallpox has been erased from the planet. The contributions of Zhdanov, Henderson, and the army of eradicators who made it happen are as impressive as the scientists and engineers who put man on the moon. Their achievement is a testament to what public health can accomplish with sufficient ambition, support, and innovation. Because of their work, nearly fifty years have passed without a single case of smallpox. Tens of millions or even hundreds of millions of lives have been saved during that time. And each year the tally grows.

Yet, close on the heels of this stellar success was a failure that could have undone years of progress. There are important lessons in that experience, too. We also must acknowledge that not all stories mark grand triumphs. There are as many failures in public health as there are in any other field: moral failures, technical failures, and failures of imagination. The progress that public health officials have fought for is fragile. We are never more than one laboratory accident, one virus spilled over from animals, one contaminated consumer product away from disaster. I offer that observation not to startle, but to reinforce that the work cannot be taken for granted for even one second. Only with great effort does public health maintain an invisible veil between us and a world that I don't want to see return.

2

COMPLACENCY

Like many old military bases in the United States, Aberdeen Proving Ground in Maryland is just off a shabby highway peppered with pawn shops and used car dealerships. Veterans will recognize the tall security fencing that snakes through the woods and behind neighborhoods as the signature thumbprint of a Department of Defense property. At Aberdeen, the fencing is adorned with signs warning that not only is trespassing prohibited but unexploded ordnance lies forgotten on the grounds.[1] Throughout both world wars it was a bustling proving ground where weapons were tested, but the site has since faded toward desolation. The base itself is split into two areas separated by a river that is frequented by fishermen willing to overlook the base's long history as a chemical weapons depot. For two years beginning in 2015, I worked as a civilian epidemiologist on the southern outpost known as Edgewood Area.

Where over 30,000 army soldiers and civilians once lived and worked, there now reside a few small research institutions focused on military public health and science.[2] The aging emptiness can be unsettling. I once found an entire neighborhood, a miniature suburbia, that had been entirely abandoned. It was well on its way to being completely overtaken by oak trees in the gutters, vines creeping up the siding, and berry bushes covering the windows. The neighborhood had apparently been built during boom times when Aberdeen was a bustling hub of activity. Then times changed. Some jobs were transferred, others eliminated. Families moved away and the place hollowed out.

It's a cycle that public health knows all too well, only we call it panic and neglect.[3] Outbreaks, pandemics, eradication programs—these events have an intensity and urgency to them that bestows energy, purpose, a sense of mission. Public health officials get right to work bringing the situation under control, and they do so with widespread support. As their efforts begin to pay off, case counts diminish. Progress, though, is not triumph. The peak of the outbreak is the halfway point, not the finish line. As transmission slows, urgency fades. The funding and attention that powered the earlier efforts wane, and progress begins to unwind. Then it's back to the beginning, and the cycle begins again. Neglect has set in. Military public health, it turns out, knows this cycle well.

ONE CRISP AUTUMN AFTERNOON, I stepped out of my office in Aberdeen for fresh air and wandered across an empty, crumbling parking lot. Tucked behind a thicket of trees at the end of the lot was a small cemetery dotted with white marble headstones. On several gravestones were the names of the deceased and the year: 1918. For epidemiologists, 1918 is an auspicious milestone in public health history. It marked the beginning of a devastating influenza pandemic that swept the globe, reaching from remote Alaskan villages across America, Europe, Asia, and all the way to the Pacific Islands.[4]

The measures that public health officials used then to slow the virus, from mandating masks and school closures to forbidding large public gatherings, formed the basis for the interventions that the world later relied upon during the COVID-19 pandemic.[5] Though prescient, these measures were not enough to stop either virus. The influenza pandemic may have killed up to 50 million people, more than died in battle across both world wars combined.[6]

For epidemiologists who work with the military, the 1918 pandemic has particular resonance. The first known cases in the United States were identified at Camp Funston located on the Fort Riley, Kansas, military reservation, an expansive training site for troops headed to fight World War I.[7] Wartime operations demanded difficult living conditions, and the winter of 1917–1918 was particularly harsh. The weather was cold, and the men were underdressed for the elements. Soldiers were stuffed together in overcrowded, under-heated barracks for just long enough to be trained up and shipped out to make way for new troops to rotate in. A virus could hardly ask for more favorable conditions—and indeed the epidemic was rampant. The first of what would be several epidemic waves swept through the camp, marking the beginning of a global pandemic.

Unlike normal seasonal influenza, which largely spares the young and healthy from severe illness or death, the 1918 pandemic hammered young adults.[8] Men who were preparing to go to war in peak physical condition were on death's doorstep days later. Dispatches from medical officers describe the horror of seeing hundreds of young men sequestered in pneumonia wards, drowning in waterlogged lungs with no treatment or means of relief. One physician wrote that within hours of admission, young soldiers would develop "mahogany spots over the cheek bones, and a few hours later you can begin to see the [c]yanosis extending from their ears and spreading all over the face . . . It is only a matter of a few hours then until death comes, and it is simply a struggle for air until they suffocate."[9] At a military camp in Massachusetts, so many men died that special trains were required to remove the dead.[10]

This pattern repeated itself at each military base in turn, as an average of 20 to 40 percent of army and navy personnel grew sick in the fall of 1918.[11] So many men were ill that the wartime effort nearly ground to a halt. This was just as well, because the constant flow of troops to the front lines in Europe is one of the factors that scholars say accelerated the virus's global spread.[12]

Military doctors recognized that their facilities, swollen with recruits drafted to fight, were tinder for the virus. They pleaded with combat leaders for relief from the relentless arrival of new personnel that overcrowded facilities. The crowded close quarters invited explosive outbreaks. But more often than not, combat leaders overruled the public health staff's requests for measures to contain the virus's spread. Camp Devens in Massachusetts, for example, became overwhelmed by 14,000 cases of influenza, nearly a quarter of its total strength.[13] At one point, the base hospital admitted 1,200 people in a single day.[14] And it was no wonder, for the camp was 10,000 men over capacity.[15] The army surgeon general asked that no men transfer in or out of Camp Devens until cases subsided.[16] It was a highly unusual request for an army at war, yet the doctors' fears were swept aside.

Similar troubles plagued the large ships that transported troops to Europe. The confined quarters and poor accommodations made easy work of transmission. Transatlantic ships arrived in Europe with thousands of passengers who fell ill on the journey and required hospitalization. On the *Louisville*, which sailed at 300 men over capacity, an outbreak of influenza infected 500 and killed 50. The *Vale* carried 2,600 passengers instead of the allotted 1,600; 45 died from influenza.[17] On some ships, as many as 2 percent of passengers died.[18] Again the army surgeon general intervened, this time calling for reduced ship capacity and warmer apparel for the men. His pleas were once again ignored.

The fate of the pandemic and the war became inexorably linked—the next installment in a long interplay between conflict and infectious diseases.[19] As Máire Connolly and David Heymann note in their review of medicine and conflict, "During the Napoleonic wars, eight times more people in the British army died from disease than from battle wounds. In the American civil war, two-thirds of the estimated 660,000 deaths of soldiers were caused by pneumonia, typhoid, dysentery, and malaria, and this death toll led to a 2-year extension of the war. These diseases became known as the 'third army.'"[20]

World War I was no exception. During the most intense periods, the pandemic all but ground military activity to a halt. Soldiers were either ill or pressed into service caring for the sick and dead, building hospital facilities, or hunkering down to avoid infection. At Camp Funston, a mild spring wave was followed by a tremendous fall surge that overwhelmed medical capacity. So many men were hospitalized that barracks, offices, and even indoor horse arenas were converted to makeshift hospitals.[21] By the end of the year, the virus infected nearly a quarter of the 63,374 personnel at the original hotspot of Fort Riley and killed nearly 1,000.[22]

Their deaths, along with the tens of thousands of other influenza deaths the military suffered that year, ushered in a new era of health awareness.[23] For decades to come, the military prepared not just for armed conflict, but for outbreaks. The panic of the 1918 pandemic had recast the military's perspective on infectious diseases and set it on course to pioneer many of the innovations that underpin modern public health—until neglect set in, and outbreaks roared back, reversing decades of progress. It's an arc that illustrates how crises can be springboards for opportunity to make progress in public health, and how quickly those gains can be lost to complacency.

FOR THE FIRST HALF of the century following the influenza pandemic, the panic phase of the cycle continued to dominate. In January 1941, with the memory of the 1918 pandemic not much faded, Brigadier General James S. Simmons was charged with preparing for the looming possibility of American involvement in the war in Europe.[24] As a young army officer, Simmons had seen firsthand the havoc wrought by the influenza pandemic.[25] Now, advising the army as the chief of the Preventive Medicine Service for the surgeon general's office, Simmons was responsible for ensuring that combat readiness was not disabled by the same explosive outbreaks that had jeopardized the Allies' success in World War I.

One of the first orders of business was controlling the "boot camp flu" that often flattened basic training classes.[26] Then, as now, respiratory infections flourished in the rough conditions endured by new recruits. At the time, it was common for a quarter or even half of each class to fall ill with fever, cough, and trouble breathing.[27] The volume of patients regularly overflowed hospitals and prompted a scramble to open new wards. At Fort Dix, nearly 50 percent of recruits were hospitalized during basic training, which one army researcher translated to over six hundred hospital admissions per week for respiratory illness at northern bases in the winter.[28] It was not uncommon for so many to be sidelined that whole cohorts had to restart training, because there were not enough recruits standing to continue the course.[29] Military leaders, ever focused on ensuring that troops are prepared to deploy, understood that these outbreaks were a serious threat to readiness. Simmons, recognizing this risk, pulled together a committee of expert scientists to defend against the "boot camp flu."

It's no wonder basic training camps are outbreak hotspots. Soldiers arrive at basic training from all over the country and they bring with them their local viral assortments. That alone would be sufficient to create the conditions for an outbreak. Worse, the recruits undergo physical and emotional stress as they are pressed into shape for military service. They run miles each day and spend hours doing physical training, often in oppressive heat or biting chill. Nutrition consists of whatever is offered at the cafeteria and is sometimes limited to prepackaged, shelf-stable meals. At night, recruits sleep in barracks in open bays with dozens of beds.[30]

Prior to the advent of vaccines, there was little that could be done to control the outbreaks except to keep troops at least notionally spread out—a task easier said than done. The military is (in)famous for being governed by endless rules and regulations. There are requirements for how eyeglasses can be styled and whether items can be carried while walking.[31] There is even guidance on proper use of pack animals. Beasts of burden reviewed include donkeys, mules, llamas, camels, dogs, and even elephants. Llamas, the Special Forces guidance notes, "have a striking beauty, owing to their elegant wool and graceful posture."[32] One (somewhat less intriguing) army rule pertains to bed spacing in barracks. The current rule specifies that each person must typically have seventy-two square feet of floor space,[33] and bunks must be situated at least three feet apart.[34] Adjacent occupants must sleep head-to-toe in an alternating arrangement.[35]

Although it is tempting to assume that this rule is meant to offer a modicum of privacy to soldiers, it is actually a preventive health measure to limit the spread of germs. Precautions of this sort date back to at least 1777, when the US Army implemented a regulation limiting six soldiers to a tent.[36] Later, during the Crimean War in the 1850s, Florence Nightingale reported that hospitals were dangerously overcrowded. Her reports led to a recommendation that each hospital bed should be allotted one thousand cubic feet of space, with each bed spaced six feet apart.[37] These observations foreshadowed the pleas of army officers during the 1918 influenza pandemic to reduce occupancy in barracks and ships to allow some degree of distancing.

The COVID-19 pandemic brought the concept of social distancing into the daily lives of the general public. Billions of people were asked to maintain six feet of space (or 1 or 1.5 meters, depending on local recommendation) to slow the spread of the virus. Shopkeepers installed floor stickers, teachers pushed apart desks, and restaurateurs removed every other chair in a bid to keep people apart. Although some researchers have asserted that the recommendation comes from early studies of influenza droplet dispersion, I suspect it originates from efforts to limit disease transmission in military settings.[38]

ALTHOUGH MILITARY PUBLIC HEALTH OFFICERS were aware of the dangers of stuffing troops together in confined quarters, in Simmons's day broader knowledge was limited. In the first half of the twentieth century, little was known about the germs that haunted the barracks. Scientists had only begun to explore the world of viruses and many pathogens we now know circulate widely were still unidentified. That changed when, over a decade after Simmons's charge to military scientists to rein in respiratory diseases in the barracks, an accidental discovery by eminent scientist Dr. Maurice Hilleman solved the mystery. In doing so, Hilleman dramatically changed military public health.[39]

Hilleman was born to a farming family in eastern Montana in 1919.[40] Originally in training to become a store manager at the department store JCPenney, Hilleman instead went on to become a leading microbiologist who invented dozens of vaccines, including nine of the vaccines on the American childhood immunization schedule.[41] In 1953, he was working at the Walter Reed Army Institute of Research (WRAIR) when he received word of an outbreak that could advance his research. Hilleman had been puzzling over why flu vaccines seemed to lose their effectiveness from year to year.[42] At the time, influenza vaccines were produced through a process involving chicken eggs. The virus used in the vaccine is injected into fertilized eggs, where it replicates into more virus. It is then harvested from the eggs for use in vaccine production. Hilleman suspected that the influenza virus was evolving in the eggs, diverging from the strain the vaccine was meant to protect against. To test his hypothesis, he would need to take samples of influenza from infected humans and compare them to egg-derived virus. When Hilleman received word of an influenza outbreak in soldiers at Fort Leonard Wood, Missouri, he felt it would be a good opportunity to advance his research.

Hilleman and his team set out at once to collect the samples. Unfortunately, their arrival in Missouri quickly led to disappointment. Hilleman found that the men's symptoms did not match those caused by influenza. This meant the soldiers had some other respiratory infection, ruining Hilleman's research plans. The mission was a costly undertaking and the realization that it was something of a boondoggle was an embarrassing blow.[43] Sheepishly, Hilleman and his team still collected swabs of the men's throats and returned home.[44]

Back in his laboratory at Walter Reed, Hilleman and his team processed the samples anyway in hopes of finding a way to salvage the work. What they found more than made up for their failure to identify influenza. The virus in the samples had never been isolated in a laboratory before.[45] The discovery of a new respiratory virus garnered Hilleman a momentous scientific discovery. (Wallace Rowe, a virologist working at the National Institutes of Health, also isolated the virus in 1953,[46] in parallel with Hilleman and his team. Each team is credited with identifying the virus independently.[47])

In subsequent years, Hilleman and his colleagues established that not only was the virus he named RI-67 responsible for the flu-like illness circulating in Fort Leonard Wood in 1953, but blood samples taken from a group of soldiers at Fort Dix showed that as many as 81 percent had been infected with the virus during a seventeen-week training period.[48] Hilleman turned a situation that began as a "blooper" into the discovery of a major cause of respiratory disease in the military.

What Hilleman called RI-67 is now known as an adenovirus, and it has gone on to enjoy a certain notoriety. Adenoviruses are a group of highly contagious viruses spread through the air, in fecal matter, and on contaminated surfaces.[49] Researchers have identified over sixty human serotypes, or variations, of adenovirus, and there may be more yet undiscovered.[50] Some subtypes are prone to causing respiratory infection while others cause gastrointestinal illness like vomiting and diarrhea, and still others cause pink eye.[51] In military settings, adenovirus types 4 and 7 are most common.[52] Although most people infected with these types develop only mild respiratory symptoms, in a minority of cases the illness can develop into a pneumonia that fills the lungs with fluid and makes it difficult for the sufferer to breathe. Outside of the military, epidemiologists still don't know how common adenovirus outbreaks are, because very little disease surveillance is conducted for adenoviruses. Outbreaks are occasionally identified in close quarters like college campuses or nursing homes, but the mostly mild nature of the disease allows it to fly under the radar, undetected.[53]

IN THE 1950S, this was just beginning to be understood. For years, military epidemiologists had been puzzling over the problem of acute respiratory disease in troops, and the motivation for doing so stretched back to 1918. Although they had made some progress in documenting the scale and nature of the problem, it was not clear what was causing the outbreaks. Hilleman's discovery unlocked an important piece of the puzzle. With the virus now identified, military scientists could devise a plan to counter it.

They did so with remarkable speed. Hilleman and his team set about creating a vaccine and by 1957, just a few years after the virus was discovered, a candidate vaccine was ready for clinical trials.[54] The results of the trial were a home run.[55] Only 5 percent of the military trainees who received the prototype vaccine ended up hospitalized, compared to 24 percent of those who received the placebo. Hilleman was elated and so was the public. A cartoon ran in the *New York Herald Tribune* depicting a syringe ejected out of an airplane as if it were a missile. Its bull's-eye target was respiratory ailments.[56]

As is often the case, the process of fully testing the new vaccine took much longer than the initial phases of research and development. It was a full fifteen years until clinical trials were complete and the Department of Defense had arranged for a manufacturer. But finally, in 1971, the vaccine was ready for use at military training camps. It was given first to male trainees starting training during the winter months, when respiratory disease outbreaks were prevalent.[57] Later, the program expanded to cover male trainees who attended basic training at any time during the year, and female trainees were added to the vaccine schedule.

As the program grew, epidemiologists marveled at the profound impact the vaccine had in reducing adenovirus outbreaks. Rates of respiratory infection in recruits plummeted dramatically and hospitalizations were significantly decreased.[58] One research team estimated that in the first two years of the program, nearly 27,000 hospitalizations were averted.[59] Military leaders, no longer faced with dozens of recruits needing to repeat their basic training each cycle, could admit more soldiers into training and shift resources away from recycling recruits who fell behind due to illness. More than that, the experience of recruits changed considerably. Prior to the introduction of the vaccine, facing a serious respiratory infection during basic training was a near certainty. After the vaccine, severe adenovirus-associated illness became uncommon.[60]

For centuries, the military had faced not one but two major foes—one in the form of combatants and the other of constant outbreaks. Now, with the military having learned from the disastrous effect the 1918 pandemic had on wartime efforts and having parried with decades of dedicated investment, one of those foes was brought to heel.

THAT SHOULD HAVE BEEN the end of adenovirus epidemics in military training, just as Ali Maalin's infection should have marked the end of the smallpox virus. But over the years, as the threat of adenovirus receded thanks to the sterling success of the vaccine, the commanders lost sight of the importance of maintaining defenses against adenovirus. Although different in form from Janet Parker's accidental smallpox infection, the neglect that led to the loss of the adenovirus vaccine was similar in spirit. Success brought complacency, and complacency brought risk. Infectious diseases do not share humanity's propensity for growing weary. Before long, adenovirus was staged to make a comeback.

In a logistical oversight uncharacteristic of the military, the adenovirus vaccine supply had always been dependent on a single manufacturer, Wyeth Pharmaceuticals. Worse, the Department of Defense was the only customer for the product. In a peculiarity that endures to this day, the vaccine is only authorized for military populations, meaning it is not available to the general public. With no other manufacturers who could make the vaccine and no market beyond the Department of Defense, the supply chain for the adenovirus vaccine was functional, but brittle. That became a problem when, as the years wore on, the facilities at Wyeth Pharmaceuticals verged on antique. In 1984, following an unflattering inspection by Food and Drug Administration officials, Wyeth informed the department that considerable investment would be needed to bring the manufacturing process up to modern standards.[61] Curiously, department officials balked at the $5 million price tag for the necessary upgrades (nearly $13 million in 2020 dollars).[62] While not a small sum, it was also not beyond the department's budget, which at the time was on the order of $250 billion.[63] In what one senior official described as a game of chicken, the department refused to produce the funds and Wyeth declined to upgrade the facilities on its own dime.[64]

In the mid-1990s, matters came to a head. Wyeth could no longer continue manufacturing the vaccine in its decrepit facility. The company notified the department that it would cease producing the adenovirus vaccine.[65] This revelation should have provoked alarm from Department of Defense officials, if not a swift return to "panic" mode.

The loss of the vaccine was an open invitation to the virus that had regularly put a significant number of a training class in the hospital. But somehow, the looming supply chain collapse did not prompt the department to change course by making alternative arrangements.

Details of what happened during that period to allow the lapse still remain murky. Military health officials overseeing the program at the time now say that they did not learn of production troubles until it was far too late to intervene.[66] It was not until the eve of the end that medical officials became aware of what had happened.[67] The decision not to fund upgrades at Wyeth seems to have resided with the part of the military charged with logistics and procurement, rather than with health.

Upon hearing the news, a group of senior army public health officials traveled to Wyeth's facilities in Pennsylvania to see what could be done to save the vaccine program. Their efforts were not well received. The officers were shown into a meeting room and settled around a conference table. The facility's director strode in, lips pursed, and dropped a large bundle of papers on the table. The stack was a history of correspondence with the Department of Defense going back a decade. Only at this late date, with the arrival of the uniformed group, did Wyeth hear of interest in maintaining the program.[68]

But of course, by then, it was too late. Wyeth Pharmaceuticals had already wound down their adenovirus vaccine manufacturing and could provide to the army only what was left on the shelves. With careful rationing, existing supplies could last a few more years. Wyeth offered to transfer the know-how to another company so that production could continue, but there was no manufacturer waiting in the wings.[69] The era of the adenovirus vaccine was coming to an end and, as epidemiologists knew all too well, a dangerous and disruptive virus lay in wait for the chance to resurge.

. . .

CYCLES OF PANIC AND NEGLECT are everywhere. In some settings, we call it procrastination. Students avoid their studies, then stay up all night to cram for an exam. Taxpayers miss the deadline to file and then prepare their submission in a frenzy when it's already overdue. At other times, we mistake panic and neglect for a sort of heroism. Congress delays passing the required annual federal budget until the eleventh hour, narrowly averting a government shutdown. All of these situations are, of course, avoidable with planning and more consistent, rather than reactive, effort.

In public health, panic and neglect are harder to spot, unless you know what to look for. Yet, like warm seas feeding hurricanes or the moon's pull on the tides, the invisible forces of panic and neglect shape so much of our work. The cycle goes something like this. A crisis hits, like the 1918 pandemic that jeopardized wartime operations. Leaders like Brigadier General Simmons resolve to give public health matters top priority, and even dedicate the resources needed to avoid whatever devastating fate was just endured. Speeches are made, laws are passed, funding flows. Often, these efforts lead to important breakthroughs. Simmons's focus on preventing respiratory diseases in troops, aided by Maurice Hilleman's good luck and immense talent, paid off in an unexpected way. A leading cause of respiratory disease was discovered, countered, and beaten back. The military went some twenty years without major outbreaks of adenovirus. In that time, tens of thousands of hospitalizations were averted.

But before long—certainly sooner than I would hope—amnesia settles like a warm blanket. Eventually, the initial conditions that allowed for outbreaks become the normal state once more. For adenovirus, the military's triumph in controlling what one army researcher called the "old nemesis" faded from memory, a victim of its own success.[70]

The enemy soon returned with a vengeance. Neglect set in, and the vaccine once heralded as a missile striking a bull's-eye was allowed to lapse. Perhaps there was some gauzy memory that priorities had blown off course. But no matter. Until—surprise! The outbreak roared back, with the same predictable ferocity as before. One cycle of panic and neglect was complete, and a new one began. Round and round it goes.

You may have witnessed cycles of panic and neglect in your own community. During the COVID-19 pandemic, governments spent billions of dollars on public health measures.[71] They set up diagnostic testing programs, installed better ventilation systems in buildings, and held regular press conferences to connect with the public. School lunch programs were expanded, and meals were given away for free. Health-care workers were loudly celebrated. But as the pandemic faded, that all dropped away. Never mind that those measures were broadly useful, not just for COVID-19 but for a range of health issues. Now, once again, most communities are back where they started in terms of readiness to face a pandemic.

In the case of adenovirus, the final stores of the vaccine were delivered to the Department of Defense in 1996.[72] Within months of ending the program, the virus again swept through the barracks of basic training camps—a specter not seen since some of the recruits' grandparents were in boot camp.

The new epidemics of adenovirus were both predictable and preventable, yet still entirely devastating. The patterns of disease and the profound impacts on basic training exactly matched what military epidemiologists had described before the vaccine was first introduced in 1971. Outbreaks swept through the camps, sickening soldiers and disrupting carefully coordinated training cycles, just as they had in Hilleman's day. Studies conducted in the post-vaccine era estimated that 10 to 25 percent of recruits became ill with adenovirus.[73] Another study calculated that if 5 percent of recruits became infected (surely an underestimate), and if 40 percent of those needed medical care, the military could expect over 10,000 preventable infections, thousands of clinic visits, and 852 hospitalizations each year for approximately 213,000 army, navy, and marine recruits.[74] The cost of related medical care for just the army was estimated at $26 million annually.[75]

Before long, tragedy struck. A young, healthy recruit arrived for boot camp at the Great Lakes training facility in Illinois.[76] One month later, he developed an upper respiratory tract infection. He visited the base's health clinic and was prescribed antibiotics for what doctors thought was bronchitis. The next day, the man was found unconscious in the barracks after complaining of extreme fatigue and blindness.[77] He was rushed to the hospital, where doctors put him on a ventilator in a bid to save his life. It was too late. Nine days later, the young recruit, just twenty-one years old, died of complications of adenovirus.

One month later, just weeks after arriving for basic training, an eighteen-year-old naval recruit developed the boot camp crud.[78] He visited the base's health clinic multiple times. On the third visit, he complained of weakness and trouble breathing. He, too, was rushed to the hospital, but just hours after being admitted, he died from acute respiratory distress. After his death, tests showed that the recruit was battling both adenovirus and a bacterial infection.

. . .

THE ABSURD, NEEDLESS TRAGEDY of the situation was not lost on observers. Commanders were facing serious losses to readiness, and news of the tragic adenovirus-related deaths was making headlines. The prestigious Institute of Medicine wrote a letter excoriating the lapse in vaccine as "extremely disconcerting" and warned that it "once again threatens the health and even the lives of military trainees."[79] *The Seattle Times* ran an article calling the situation a "major screw-up."[80] Pressure mounted on military leaders to account for the situation.

The Department of Defense finally took action. Seventeen years after Wyeth Pharmaceuticals requested funds to upgrade their facilities, and five years after production of the vaccine ceased, the Department of Defense reversed course. On September 26, 2001, the US Army Medical Research Acquisition Activity awarded a new contract to re-establish production of the adenovirus vaccine.[81] But it was not so simple as hiring a new manufacturer. Although the department could have invested some $5 million to secure a continued supply had it done so in the 1980s, it now faced an expensive gauntlet that included establishing production facilities, conducting a new round of costly clinical trials, and undergoing inspection and certification by the Food and Drug Administration.[82] One review of the acquisition process for the new vaccine tallied that over two hundred personnel were involved in the process. The effort cost around $100 million and spanned a decade, from 2001 to 2011.[83] During that period, there was no adenovirus vaccine available to protect trainees. In the interim, six more deaths followed, bringing the total to eight.[84] All were young recruits just beginning their careers. Until then, the military had not recorded any deaths from adenovirus since 1974.[85]

Any time a young person dies, the pain of a life cut too short is anguishing. A stray bullet. A fatal tumor. A car crash. These twists of fate resist efforts at consolation and shatter loved ones with grief. The loss of eight young people who volunteered their lives in service of their country and lost them to a preventable disease instead is something else entirely. Their deaths, in all likelihood, would have never happened had the well-known, proven-effective adenovirus vaccine remained available as it had for decades. But neglect set in. Over the years, the scourge of adenovirus epidemics became a distant memory. Military health professionals knew of the threat in abstract, but they had personally witnessed only the *absence* of disruption. I suspect that this remove did not equip them with the urgency needed to maintain defenses.

DESPITE BEING ONE of the plainest examples of the vicious cycle of panic and neglect in public health, the problem of adenovirus in the military is at least a problem that was solved. (Twice.) Notwithstanding the lives of eight young recruits, the department was able to set things right for a tidy $100 million. Although restoring the adenovirus vaccine supply was not easy or cheap, it was fairly straightforward. The department followed a series of well-defined steps: choose a new manufacturer from among the limited set of qualified candidates, issue a contract through existing procurement mechanisms, and follow the regulatory process to obtain needed approvals. The loss of the adenovirus vaccine was a problem for which the path to a solution was clear. For some other public health matters, resolving neglect is a long time coming, if it happens at all.

To say it plainly, cycles of panic and neglect cannot be allowed to govern public health. The risks are too great, and the costs are too high when neglect inevitably gives way to new (or renewed) reasons to panic. Public health can do so much to improve health and well-being in our communities. The eradication of smallpox changed the course of history, and the adenovirus vaccine fundamentally changed boot camp. But it can never do so with the constant setbacks born of neglect. And neither, for that matter, is panic desirable. What we need most of all is the steady march of progress that comes from sustained support. When vaccines are discontinued or made optional, complain. When funding for your local public health department is cut, protest. When it's time for your annual flu shot, go. When public health calls, answer.

3

SKILLS

Every student epidemiologist learns the origin story of the founding of the field. Its hero is Dr. John Snow, an English physician who lived during a heady time in science and medicine, alongside the likes of William Farr, Florence Nightingale, and Louis Pasteur. Among these seminal figures, Snow holds his own in the history books. By the time he reached early middle age, Snow had already achieved remarkable renown in multiple fields. His earliest fame came from his work with chloroform, a sweet-smelling gas that renders a person unconscious when inhaled.[1] Prior to the advent of chloroform, surgeons had little to offer in the way of pain relief—other than to cut, saw, and sew as quickly as possible. Often, the pain was so tremendous that patients simply passed out. They were the lucky ones. Snow and his contemporaries put chloroform to merciful use as a surgical anesthetic, making him one of the first anesthesiologists.

Snow's busy medical practice took him across the highs and lows of London society. He traversed the city's streets, aiding surgeons and dentists around town, bringing relief to laboring women at home, and lecturing at the prestigious medical societies of the day. He even eased the pain of Queen Victoria as she delivered her eighth child, a milestone that helped pave the way for acceptance of the practice; labor pains had previously been regarded as God's will.[2]

The same blend of courage, vision, and careful study that powered his success in anesthesiology led him to investigate the origins of cholera. Cholera has a prominent history in the annals of public health. It is a capricious disease, with only some people developing symptoms, an attribute that likely muddied the path to its discovery. For centuries, the origins of cholera were regarded as supernatural and, often, punitive. Prevailing theories included disfavor from the gods or exposure to miasma, or "bad air."[3] Most people who become infected experience either no symptoms at all or a mild bout of diarrhea. But for a minority of victims, the bacteria cause catastrophic diarrhea so incessant that it can kill in a matter of hours.[4]

In Snow's time, cholera broke out regularly in London. Snow trialed chloroform as a potential cure, but quickly found it to be useless against the infection. And effective treatment was not the only open question. Scientists at the time were still trying to work out how cholera spread. Many "medical men" at the time debated over whether cholera was spread through miasma or through blood.[5] Still others suspected a food as a source. The truth remained an open question for years, vexing scientists and driving lively debate in the medical journals and supper clubs.[6]

The problem wasn't just that scientific understanding was primitive. Even now, pinpointing modes of transmission is difficult. People who have been in contact with one another often share multiple forms of contact. They may socialize within breathing distance, drink from the same water source, eat the same foods, ride the same bus, touch the same surfaces, or receive bites from the same bugs. Any of these activities could transmit disease. Matching the pathogen to the specific activity requires extensive, careful study—and a bit of luck. On these merits, Snow excelled. He devised a series of elegant investigations that earned him a place in the history books.

Snow's approach was methodical: he collected and analyzed data collected from the cholera-affected Golden Square neighborhood, inquiring about whether each household had been stricken by cholera. He also asked about residents' workplaces and the sources of their drinking water. Snow recorded the data he collected from each interview on a hand-drawn map of the district.[7] Each instance of cholera was represented by a small black square on the map. In areas where the disease was more prevalent, these squares clustered into little towers. He also marked the locations of water pumps, a key focus of his investigation, with black circles.[8]

This careful application of scientific methods revealed a clear pattern: most cholera cases were found in households near the Broad Street water pump, implicating it as a likely source. But perhaps the strongest evidence came from outlying cases farther away. Snow's research revealed that some people preferred the water from Broad Street and went out of their way to fetch it. Other distant cases worked at a factory that drew water from the pump. These findings, along with the detailed map, were conclusive. Water from the Broad Street pump, not bad air, was the cause of the Golden Square cholera outbreak.

What many people do not know is that John Snow's now famous map was just one of a series of analyses he did to establish contaminated water as the source of cholera outbreaks. He worked fastidiously for years to collect data on the rates of cholera in households served by various sources of water. One of his close intellectual partners was William Farr, a fellow pioneer in epidemiology who is now considered the founding father of medical statistics. Farr was the first to recognize that during epidemics, the number of new cases rises and then falls in a now-familiar bell curve shape.[9] Together, Snow and Farr made a formidable team. Their elegant work to use statistical methods to discover the source of cholera both solved several important mysteries and founded what we now know as infectious disease epidemiology.[10]

THE INVESTIGATIVE approach that Snow took still forms the basis of modern disease detective work. Epidemiology is defined in most textbooks as the study of the distribution and determinants of disease in a population, which aligns with Snow's objectives.[11] Then, as now, epidemiologists collect data on the who, when, and where of diseases in hopes of finding a pattern that will reveal how a pathogen spreads and how it can be stopped.

Many modern outbreak investigations begin with interviews of the people affected. This process is more thorough than what the terms "case investigation" or "contact tracing" imply.[12] Epidemiologists ask detailed questions about when someone became ill, who they had interacted with, and everything they did in the days and weeks before their illness. Depending on the illness, interviewees may be asked about their diet, water sources, and mundane habits. For complex or mysterious outbreaks, the interview can involve hundreds of questions. These data are compiled into a spreadsheet called a "line list," which is used to uncover patterns like the one that led Snow to the Broad Street pump.[13]

Many scientific disciplines have evolved toward favoring "big data," or sprawling datasets that contain enormous quantities of data on everything and anything. The large language models that power current advances in artificial intelligence like ChatGPT are trained on practically the entire internet. Such big data approaches have been trialed in epidemiology. Data scientists have built epidemiological models that call on, for example, cough syrup purchases, internet search history, biometric readings from smart watches, and so on.[14] In my opinion, these approaches will always supplement, not replace, traditional outbreak investigations. The big data models can't give specifics about who is infected, where the problem is concentrated, or what cases have in common. Without those details, there's not much for epidemiologists to do other than to admire the problem. In my experience, quality trumps quantity. When I am asked what data are essential to investigating emerging outbreaks, my answer is always: "a line list."

For the most important disease detective work, all that is needed is a trained investigator, a notebook or tablet, and a calculator. None of this is as complicated as, say, experimental physics. But don't let the low-tech simplicity fool you. The power of collecting and collating this information is extraordinary. Take a recent example. In 2021, a handful of cases of a rare bacterial infection called melioidosis were diagnosed, one each in Georgia, Kansas, Texas, and Minnesota.[15] Melioidosis occurs naturally in water and soil in some parts of the world, but rarely in the United States. When sporadic cases are diagnosed, often they are found in the Gulf Coast region.[16] None of the cases in 2021 lived in that area, nor had they traveled or had contact with one another. The source of the outbreak was a mystery that had public health officials stumped.

It was only through an exhaustive investigation that epidemiologists cracked the case. The culprit was a most unexpected source: an aromatherapy room spray sold at Walmart.[17] Melioidosis had never been found in a commercial product before, and certainly not in an air freshener. The surprising conclusion was reached through the same type of epidemiological investigation that Snow used to link cholera infections to a specific Broad Street pump. The product was pulled from store shelves and consumers were warned to throw away any bottles at home, ending the outbreak.

Or take another, theoretical, example. If I learn that one third of a preschool class has stayed home with fever, vomiting, and fatigue over the course of three weeks in February, and that numerous family members are showing similar symptoms, I'm not particularly concerned. That's typical for that age group, at that time of year. (As a mother of young children, I can personally attest.) If you add the details that they all have a rash and stiff neck and four are hospitalized, we're going into emergency response mode, because that sounds like bacterial meningitis.

In modern times, Snow-style epidemiological investigations are supplemented with microbiological and genomic analyses. Biological samples are taken from the patient and collected from possible sources of exposure that are under investigation, like contaminated water or food. The specimens are analyzed in a laboratory to identify the responsible pathogens. Often, the pathogen's genetic code is sequenced for more detailed examination, including looking for mutations or clues that can help to develop effective treatments or vaccines. These approaches are helping outbreak scientists to learn even more about how outbreaks begin, how they spread, and who is at risk.

These investigations are not for scientific edification. The objective is to identify the places where public health interventions can break the cycle of transmission and slow or halt an epidemic. In Snow's case, he is often credited with using his analysis to convince city officials to remove the pump handle that was used to draw water from the Broad Street well.[18] In popular retellings, it is said that this decisive stand stopped the epidemic in its tracks, ending the Golden Square cholera outbreak. In truth, the outbreak was already waning before Snow concluded his investigation, likely in part because residents decamped from the affected area. But regardless, Snow's skills as a careful scientist and astute observer of patterns established the capacity of epidemiology to reveal the hidden patterns that drive epidemics.

Over 150 years later, these same basic steps enabled scientists to recognize, understand, and combat one of the most serious would-be public health crises of the early twenty-first century. The outbreak stands out as a hair-raising close call—and one of the most extraordinary demonstrations of skills in epidemiological history.

In February of 2003, Dr. Liu Jianlun faced a challenging stretch at work. Patients with a devastating pneumonia were showing up to the hospital in Guangdong, China, where he worked as a specialist. The pneumonia waterlogging his patients' breathing was difficult to diagnose and treat. Chest X-rays revealed lungs that looked like they were filled with ground glass. Many of the medical staff at the hospital had become infected, too, a pattern that makes epidemiologists' blood run cold. Doctors and nurses are accustomed to safely caring for people without becoming ill themselves. When those basic protections fail, it often heralds trouble.

Liu himself had been recently troubled by pneumonia, but after a course of antibiotics he felt well enough to travel three hours by bus to Hong Kong for his nephew's wedding.[19] Upon arriving on February 21, Liu and his wife enjoyed an afternoon of shopping and dining with relatives before retiring to the three-star Metropole Hotel.[20]

Overnight, Liu's pneumonia took a turn for the worse and he grew increasingly feverish. By morning, it was clear that something was seriously wrong. Liu and his wife checked out of the hotel and went to nearby Kwong Wah Hospital, where he was admitted to intensive care with fever and difficulty breathing—symptoms eerily similar to his patients he had contact with in Guangdong who were showing up to the hospital unaccountably ill.[21]

Liu never recovered from his illness. He died on March 4, less than two weeks after arriving in Hong Kong, nearly all of which was spent in critical condition.

Shortly before his death, more troubling developments emerged. Liu's brother-in-law was admitted to Kwong Wah Hospital with the same severe symptoms.[22] His sister also fell ill, as did his wife and daughter who by then had returned home to mainland China. More cases soon followed.[23]

ALTHOUGH THE EVENTS were not reconstructed until much later, Liu's stay in Metropole Hotel accelerated the mysterious virus's leap from Guangdong to Hong Kong and on to other countries around the world. Epidemiologists later tallied at least sixteen other guests at Metropole Hotel who became infected with the virus, including seven who stayed on the same floor as Liu and his wife.[24] Many of those guests unwittingly carried the virus to their next destination, where new outbreaks took root.

One Metropole Hotel guest, a Chinese American businessman named Johnny Chen, stayed across the hall from Liu before continuing on to Vietnam.[25] Like Liu, Chen fell sick with a fever, fatigue, and cough shortly after arriving in Vietnam.[26] Chen sought help at the Hanoi French Hospital,[27] where doctors, unaware of what was unfolding in China and Hong Kong, suspected he was infected with avian influenza. Such cases are considered serious public health events because influenza has long been regarded as a top pandemic risk.[28]

Chen's illness concerned his doctors enough that they contacted Dr. Carlo Urbani, a World Health Organization official stationed in Vietnam.[29] Urbani was an Italian physician-epidemiologist whose career had already taken him to posts in the Maldives, Cambodia, and the Philippines. Although his expertise was in parasitic diseases, he had a reputation as a top-notch diagnostician.[30] Perhaps he would be able to find what was sickening the businessman.

Urbani examined Chen and was troubled by his rapid health decline. Chen, previously young and healthy, was now critically ill with respiratory failure. Moreover, in the days that followed, twelve health-care workers who treated Chen were also admitted to the hospital with severe respiratory issues, which was evidence that the disease was spreading easily. One doctor or nurse falling ill could be regarded as bad luck, or a lapse in infection controls.[31] A dozen was a warning sign. Urbani realized that whatever was affecting Chen was highly dangerous—a pathogen that was deadly, infectious, and possibly brand-new. He consulted with trusted colleagues in Vietnam and the Philippines about his concerns.[32] The epidemiologists wondered: Could Chen's illness be related to rumors of an outbreak of pneumonia of unknown cause in Guangdong?

Sensing a burgeoning crisis, the team alerted Vietnamese and international health officials of their fears that a dangerous epidemic was brewing. The experts had no way of knowing just yet that it was a novel virus circulating, and it would take several more months for scientists to identify the exact cause.[33] They also did not know just how deadly or easily transmitted the virus would turn out to be. But even without these answers, Urbani and his colleagues had an uncanny intuition that what they were seeing in Vietnam was worth raising an alarm.

The epidemiologists were right. The spread of pneumonia from Guangdong to Hong Kong and then to Vietnam marked the beginning of a global epidemic.[34] Although the cases involving Liu and Chen were not the first infections of what was later named SARS-CoV, or severe acute respiratory syndrome coronavirus, they were the first recognized cases in a cluster that eventually made SARS a household name.[35]

Urbani's warning led to a World Health Organization alert that prompted clinicians and public health officials worldwide to hunt for cases in their own regions, suspecting that some had likely gone undetected.[36] Crucially, the alert came early in the outbreak, before the virus gained a strong foothold. Once a virus becomes too entrenched, it becomes wickedly difficult to bring under control. Urbani's actions may have been the difference between humanity's victory over the outbreak, and the SARS virus spinning out of control.

Tragically, Dr. Carlos Urbani never got to see the events he set in motion.[37] He became infected with SARS-CoV during his investigations and died from the disease. He was survived by his wife and three children. He was just forty-six.

WHEN I THINK about the phases of an outbreak and the race to gain the upper hand, so much hinges on the skills of the experts rushing to respond. At each phase, there are skills both learned and intuited that spell the difference between success—bringing the outbreak under control—and failure. The first step of a successful outbreak response is detection or identifying that something is amiss. Urbani was gifted as an "astute clinician," a term used to describe the ability of skilled health-care providers to sense that something is wrong. Sometimes, they can't even put their finger on what they see that is bothering them. But they learn to trust their instincts and act on those suspicions. Around the world, doctors are encouraged to report any odd findings to local public health authorities, and the United States Centers for Disease Control and Prevention runs a 24/7 hotline for receiving such reports.[38] Even in countries with technologically advanced disease surveillance systems, it is often astute clinicians who uncover early outbreaks.[39]

Physicians rely on their skills in detecting patterns to practice their craft. They hunt for patterns in the constellation of symptoms, medical history, and laboratory or radiological results that make up a patient's care. Each of these observations is sorted, matched to possible diagnoses, with sometimes only the subtlest of details differentiating one possibility from another. And once a diagnosis is made, physicians draw on a wellspring of learned and intuited patterns to select the best treatment plans for the patient in their care. Urbani must have seen Chen, previously healthy and now seriously ill with an infection, as one red flag. When other health-care workers at the Hanoi French Hospital began falling ill, that was red flag number two. Together with rumors circulating of an outbreak in China, the pattern pointed to something both severe and transmissible.

When a new disease appears, as with SARS, it is notable not for what it *is* but for what it is *not*. For a doctor to recognize that they just are faced with patterns not only new to them but to the human population as a whole, they must be extraordinarily skilled and observant. They must be able to eliminate all other candidate diagnoses and be left with an ominous possibility—a new virus has emerged.

Even as Urbani was examining Chen, the virus continued racing around the world. Another Metropole Hotel guest staying a few doors down from Liu, seventy-eight-year-old Sui-chu Kwan, traveled from Hong Kong to her home in Toronto.[40] Soon after arriving, she fell ill and died of a presumed heart attack, which at the time was thought to be related to her chronic health conditions. Her death did not raise suspicions until days later, when her adult son also fell ill and sought help at the local hospital. Others soon followed. Kwan's son was one of at least four family members who developed symptoms.[41] The cluster touched off chains of transmission that fanned through Toronto hospitals, sending the province of Ontario into a state of emergency.[42]

As Toronto grappled with its outbreak, Singapore was careening down a parallel path. Three travelers who stayed down the hall from Liu at the Metropole Hotel were admitted to Singapore hospitals. One young woman unwittingly spread the virus to at least twenty-one others, mostly health-care workers.[43] Among those she infected was a thirty-two-year-old doctor, Hoe Nam Leong, who developed what he believed to be dengue fever.[44] As Leong began to recover, he felt well enough to continue with a planned visit to New York City, together with his wife and mother-in-law.[45] But while there, Leong's symptoms returned with a vengeance, and he was wracked with chills and pain. He phoned a colleague back in Singapore with an update that he had been diagnosed with atypical pneumonia before he boarded the first flight home.

Unbeknownst to Leong, the call set off a flurry of events. His colleague in Singapore realized that Leong, now in the air, was likely infected with the mysterious new disease, and contacted health officials. News that there was a seriously ill doctor flying over the Atlantic Ocean along with more than three hundred other passengers was forwarded to country health officials and then the World Health Organization Global Outbreak Alert and Response Network (GOARN) team. The team had just a few hours to plan a high-stakes international operation. By the time Leong's plane landed in Frankfurt to refuel, everything was in place. Three German infection control experts in "spacesuits" boarded the plane and whisked the family to an academic hospital, where all three were treated in specially equipped isolation units until they recovered.[46] The mad scramble to arrange their care was just one of many extraordinary measures that unfolded as the world came to grips with the SARS pandemic.

To this day, for all the attention and investigation into the SARS outbreak and its global spread, it remains a mystery how guests became infected while at the Metropole Hotel. No direct interactions between the other patients and Liu Jianlun were ever identified. Some epidemiologists speculate that Liu may have vomited outside his hotel room, and that the subsequent cleaning may have lofted virus-laden particles into the air.[47] Whatever the route, the burst of cases that emerged from the Metropole Hotel upgraded the SARS outbreak to a worldwide pandemic.

DURING THE EARLY STAGES OF an outbreak, epidemiologists work quickly to understand the behavior of the virus. The SARS virus was notable for its unusually high mortality rate. COVID-19 killed an estimated 1 percent of people infected prior to availability of the vaccine.[48] The SARS virus, in contrast, had a much higher fatality rate at 11 percent, meaning approximately one in ten victims died.[49]

This estimate of deadliness is one of the most important questions that epidemiologists must answer when investigating a new or emerging outbreak, because it guides decisions about resource allocation and control measures. A virus that is very lethal justifies aggressive control measures, while a virus that causes only mild symptoms merits a more measured approach. Marburg virus, for example, has killed up to 88 percent of reported cases in prior outbreaks.[50] Even a small outbreak of Marburg could strain hospitals because they would need to care for critically ill patients while ensuring fastidious infection control to protect staff. The viruses that cause the common cold, on the other hand, do not receive much attention other than general-purpose reminders to wash hands and cover your cough because they are typically quite mild.

Estimating lethality is straightforward, in principle. The case fatality risk (sometimes also called the case fatality rate and known more colloquially as the mortality rate) is simply the number of deaths divided by the total number of cases. But early in an outbreak, this quick calculation is subject to several pitfalls that can lead public health authorities astray. The troubles begin right from the start. In order to be included in the counts, cases must have some reason to come to the attention of epidemiologists. Often, this happens because a doctor like Urbani has one or more patients who are unexplainably very ill. By virtue of the cases being detected in this way, early estimates of severity are often falsely high, because both the numerator (the number of deaths) and the denominator (the number of known cases) will mostly comprise cases who were identified *because* they are quite sick. It's as if you were tasked to estimate how tall the average human is by using only a basketball team. Your guess would inevitably be wildly wrong.

As the outbreak grows, though, the opposite problem emerges. As the virus spreads, more and more people are newly ill. Those cases are added to the denominator, and so the ratio of deaths to cases begins to look more favorable. But this is misleading. It can take days, weeks, or even months for someone who is battling an infection to recover or die. During that time, it's unclear what their outcome will be. Careful adjustments to the case fatality risk must be made to account for these delays.

Laboratorians are part of the early investigative team, too. The Metropole Hotel episode recalled a famous outbreak in 1976, when over two hundred men from across Pennsylvania developed a mysterious infection marked by high fevers and pneumonia.[51] The victims were mostly military veterans who had recently attended an American Legion convention held at a hotel in Philadelphia. The hotel, the Bellevue-Stratford, bears an address of some irony to epidemiologists: 200 S. Broad Street, recollecting John Snow's study of cholera on London's Broad Street.

Although the convention was clearly a common thread, doctors struggled to identify what was making the men sick.[52] Dozens of tests failed to pinpoint a cause: it was not influenza, it was not heavy metal poisoning, and it was not a poison gas causing their symptoms. The twenty epidemiologists sent to investigate the outbreak were at a loss. It wasn't until months after the convention, when a self-described "backbencher" microbiologist, Joseph McDade, then thirty-six, finally narrowed in on an answer.[53] McDade's job was to rule out rickettsial infection, a kind of bacteria. McDade checked the samples, found no rickettsia, and called it a day.

Then, a chance encounter changed history. In a 2016 interview with radio station WHYY, McDade described a conversation that nudged him to take a second look. While attending a Christmas party, a stranger approached: "He said, 'I know some of you scientists are sort of, kind of strange, but we count on you when these sorts of things come up, that you're able to figure these things out,'" McDade told WHYY. "He went on for a little while. His tone of his voice was very clear: he was very, very disappointed."[54]

McDade went back to the lab.

There, he found a rod-shaped bacterium that causes what we now call Legionnaires' disease. It is an uncommon infection typically spread through aerosolized water, like the fine mist produced by HVAC systems, Jacuzzis, or fountains.[55] Had it not been for the uncomfortable Christmas party, the outbreak may have remained undiagnosed.

But it was not *Legionella,* which does not spread from person to person, that lurked in the Metropole Hotel. Laboratory scientists discovered that it was a new virus from the family of coronaviruses that, at that time, were known to cause only common colds. This virus produced much more than a cold—patients with SARS were dying of severe pneumonia, their lungs so filled with fluid that they were unable to get enough oxygen even when intubated. It was the events of 2003 that introduced coronaviruses as a new class of serious, pandemic-prone respiratory viruses, alongside the likes of influenza. There are now three coronaviruses known to cause severe illness: SARS-CoV, SARS-CoV-2, and MERS-CoV, which is found mostly in the Middle East.[56]

FOR THOSE WHO LIVED through the COVID-19 pandemic less than two decades later, the chronology of the global response to the SARS pandemic will feel familiar. After percolating in mainland China and surfacing in Hong Kong, Vietnam, Singapore, and Canada, the virus zipped effortlessly around the world, carried by unwitting travelers.

In the Greater China region, the impact was greatest, reaching almost 7,500 cases. Throughout the rest of the world, the virus eventually reached at least twenty-five other countries, with case counts in some places reaching into the hundreds.[57] But for all that the 2003 SARS pandemic looks to contemporary eyes like a precursor to the

COVID-19 pandemic, there is one key difference—despite chasing a virus that had already reached thousands of people across dozens of countries, epidemiologists succeeded in bringing it to heel. What's more, they did so without the help of a vaccine or easily available diagnostic testing. The approach epidemiologists took is what could be called the old-fashioned way: contact tracing, isolation, quarantine, and instructions to the well for how to prevent infection. These skills have been used and taught by public health workers for generations.

In the case of SARS, hospitals emerged as epicenters of transmission, and so many of the infection control measures meant to contain the pandemic were targeted there. Patients with SARS were most contagious during the later stages of their illness, when they needed hospitalization. Medical procedures like intubation posed significant transmission risks to the health-care workers at the bedside. To manage the risks, patients thought to be infected were treated in specially designed hospital rooms built to handle patients with dangerous infections. In some cities, wards or even entire hospitals were set aside to care exclusively for suspected patients.[58] Doctors and nurses donned personal protective equipment like gowns and masks before entering patients' rooms, and only those trained in infection control were assigned to care for SARS patients. These procedures, though onerous for clinicians and lonely for patients, protected staff and minimized the chance of the virus spreading.

Meanwhile, epidemiologists undertook the disease detective work needed to identify new cases and control the virus's spread. They conducted contact tracing by interviewing each person who was infected, creating a picture of the network of people each sick person had exposed and so might next become sick.[59] Each of those contacts was then asked to quarantine by avoiding all physical closeness with others. While quarantine cannot prevent someone who has been exposed from becoming sick, it does ensure that the virus does not get passed on further by ending chains of transmission.

Contact tracing is very effective if outbreaks are small enough for the investigators to reach everyone who might have been exposed, and it's used regularly to manage diseases like Ebola, tuberculosis, and sexually transmitted infections.[60] But there are limitations. If an outbreak grows too large, too quickly, it can outpace the ability of tracers to keep up.[61] Moreover, the process of contact tracing and arranging isolation or quarantine can take many hours, and it must be done very quickly. In order to be effective, contract tracers must reach contacts at risk within a few days of their exposure. Their work must be thorough, fast, and sensitive to the stress that comes with being sequestered, to say nothing of being told you might have been exposed to a deadly disease. Contact tracing is also logistically complex. People who are isolated or quarantined for respiratory diseases are not allowed to go to work, the grocery store, or the laundromat. Public health teams must arrange for the provision of basic needs like food and medication delivery. Sometimes public health officials even arrange for a hotel room so roommates or family members are protected. During the COVID-19 pandemic, New York City made 11,000 hotel rooms available, free of charge, to people who needed to isolate away from home.[62] In Seattle, over 1,300 hotel rooms were provided for the unhoused to keep them out of crowded shelters and encampments.[63]

CONTACT TRACING ALSO HELPED epidemiologists reconstruct transmission networks, which revealed an unusual property of the SARS virus—a revelation they used to their advantage. They discovered that SARS was especially likely to spread in crowded indoor settings like weddings or restaurants. This characteristic turned such gatherings into "superspreading events," where a single convening could result in many infections. Superspreading events are enormously dangerous, sometimes claiming dozens of victims. In addition to hospitals, religious gatherings, shared car rides, and residences were also implicated in several large clusters.[64] The Metropole Hotel was linked to sixteen cases, and a single large apartment building in Hong Kong was linked to over three hundred cases.[65] A thorough investigation of the apartment building revealed that faulty plumbing in the form of an ineffective drainpipe seal allowed the virus to spread from floor to floor through the building's pipes.[66]

There is a silver lining when it comes to combating a virus prone to superspreading. Despite its sharp edge, an outbreak driven by superspreading events focuses the work of contact tracing. The nature of superspreading suggests that if one person was infected at an event, it's very likely that others were, too.

Consider, for instance, a situation where someone is diagnosed with SARS. Upon investigation, epidemiologists determine that the person likely became infected while attending a birthday party four days prior. Contact tracers should prioritize reaching out to other party attendees, because if there was one transmission at the event, there were likely others as well. This targeted approach in tracing and quarantining contacts from a superspreading event is more efficient than dealing with many isolated cases spread out over a wide area because there's also likely to be a list of attendees and, if the tracer is lucky, contact information.

There's another silver lining, too. Superspreading is often offset by limited spread in other situations. With SARS, for example, *most* cases did not transmit the virus to anyone else.[67] This means that contact tracers do not have to be perfect in their efforts to stay ahead of the virus. If they fail to interview a few cases or miss a few contacts, the odds are that the virus won't spread further—unless, of course, the missed case touches off a superspreading event. Put another way, epidemiologists can miss some cases here and there, but they cannot afford to miss a superspreading event.

This uneven distribution of transmission, sometimes called the Pareto principle or the 80/20 rule, governs many natural phenomena, including the spread of some diseases. In the context of epidemiology, the principle suggests that a significant portion of disease transmission—around 80 percent—stems from a relatively small number of cases, approximately 20 percent.[68] Epidemiologists have recognized and leveraged this pattern in their efforts to stop outbreaks. By focusing on identifying and managing these key superspreading events, they can effectively cut off a large portion of potential transmission.

WITH OUTBREAKS, time passing is often an enemy, not an ally. In cities around the world, epidemiologists struggled to get ahead of SARS as the virus was imported by travelers and then ripped through networks of health-care workers, family members of the ill, and sometimes even people for whom no link could be found. For weeks, case counts grew day over day, sometimes clearing one hundred new diagnoses.[69] Officials worked at a breakneck pace.[70] The World Health Organization held teleconference calls daily for epidemiologists around the world to share news and reports. Laboratorians, most of whom normally worked on influenza, cobbled together a working group devoted to testing for the virus. Health-care workers treating infected patients moved into hotels or sent their kids to live with relatives in an attempt to keep their families safe.

In Hong Kong, the outbreak intensified week over week after Liu checked into the Metropole Hotel.[71] A series of superspreading events generated hundreds of cases, in short order threatening to overwhelm the capacity of the public health and medical systems. In China, where the outbreak started, early denials by officials transformed into an all-hands-on-deck approach to stopping transmission, with the Chinese president Hu Jintao declaring a "people's war" on the virus.[72] In Canada, the Toronto outbreak ballooned into the worst outside of Asia, eventually reaching 251 cases.[73] So far-reaching was the crisis that non-emergency medical care at Toronto hospitals was halted and thousands of people in the city were quarantined.[74] In one affected hospital, 22 percent of emergency room staff and 60 percent of cardiac unit staff were infected, a devastating testament to how easily the virus could spread in hospitals.[75]

Outbreaks have an intensity and urgency to them that bestows energy, purpose, and a sense of mission. Public health officials get right to work bringing the outbreak under control. But the period after an outbreak begins to recede can bring challenges of its own. Efforts begin to pay off and case counts diminish. Progress, though, is not triumph. The peak of the outbreak is the halfway point, not the finish line. But as case counts begin to fall, so too do the funding and attention that powered the earlier efforts. For public health officials, the last leg of an outbreak often must be tackled more or less alone. It's the dangerous moment in the cycle of panic and neglect, the scourge of complacency, played out here in the life cycle of an outbreak.

The epidemiologists pursuing SARS persisted nonetheless, perhaps daunted but never conceding defeat. Epidemiologists in dozens of countries doubled down on tracking and isolating new cases, each inspiring a new round of contact tracing. They kept their sights on the goal of containing the dangerous new virus by methodically identifying each and every chain of transmission and moving to cut it off before it could grow new links. Eventually the persistence paid off, and epidemiologists began to gain ground. Daily case counts began to slow, and then decline, as transmission was interrupted and the virus was blocked from finding new victims.

But even that was not a sufficient victory. Epidemiologists did not surrender to complacency but pursued the virus doggedly until it was fully extinct. Finally, after 9 months, 8,096 confirmed cases, and 774 deaths, SARS was completely eliminated from the human population.[76]

THE END OF THE OUTBREAK, though triumphant, did not mark the end of the work. There was still the matter of discovering where the virus came from in the first place. Just because there are no human cases, that doesn't mean SARS has vanished from the planet—or that it will never resurface again.

From the earliest days of the outbreak, it was widely assumed that the virus came from animals. Most pathogens do. Contact with wild animals, agricultural animals like cows and pigs, and even domestic pets are all common ways that humans encounter viruses and bacteria. These animal-human transfers of disease are known as zoonoses. One team of researchers estimated that over 60 percent of pathogens that infect humans are zoonotic, and for new pathogens it's more like three in four.[77] Remember that next time you visit the petting zoo.

It was not a surprise, then, that a closer look at the earliest human SARS cases in China revealed that frequent contact with animals was a common theme. Investigators found that many of the first known SARS cases in Guangdong worked with animals, particularly at markets where live animals were sold and slaughtered for food.[78] Scientists even discovered a virus similar to SARS-CoV in several animals sold at the markets, including Himalayan palm civets and a raccoon dog, while a Chinese ferret badger exhibited the presence of SARS-CoV antibodies.[79]

Whichever animal was the bridge to humans, it was likely an intermediate stop and not the virus's primary host. That mantle almost certainly belongs to bats, which carry an astonishing array of viruses. Indeed, the role of human contact with wild animals, and bats specifically, is one of the strongest patterns in how, where, and when infectious diseases emerge.[80] Bats also carry the viruses that cause Ebola, rabies, Nipah, and a range of other nasty infections. Whether through direct contact with bats or through an intermediate host, the tiny mammals play an outsized role in seeding human outbreaks. Even now, two decades later, this thin sketch—bat to animal (perhaps civets?) then on to humans—is more or less the sum total of our understanding about the origins of SARS-CoV.[81] I hope fervently that we never have a need to learn more.

Regardless of its origins, the public health response to the next epidemic will likely use the same playbook as the SARS response, because those same tools and strategies remain our mainstays. Although the field of epidemiology has advanced markedly since John Snow first illustrated his cholera map, we rely on many of the same tools and approaches that he and his contemporaries pioneered. Epidemiologists still interview cases and their families to collect data about where they may have been exposed and analyze that data for patterns that can reveal the source of a mysterious illness. Although we have added new tools to the arsenal over the years, including microbiological analyses and more sophisticated forms of disease surveillance, the essential skills of the field endure.

4

FOUNDATIONS

If there is one thing that I find miraculous about public health, it's the ability to transform something into nothing. Smallpox, gone. Diphtheria, forgotten. HIV not yet banished but at least defanged. Making specters disappear is something of a specialty. What we can't do, though, is go the other way. We can't make something out of nothing. We are scientists, after all, not magicians. We need infrastructure—supplies, money, staff—to work our alchemy.

Academics and practitioners know this, of course. Students of public health learn of a concept known as *determinants of health*.[1] It refers to the factors in our lives that influence or determine our health. Many are outside our direct control. Genetics, for example, is a determinant of health. If your parents lived to 101, your odds of seeing old age are pretty good. But you don't get to choose your parents and you don't get to choose your genes, so genetics is considered an immutable, or unchangeable, determinant of health. Occupation is another determinant, for myriad reasons. Someone who sits all day at work may struggle with cardiovascular fitness, while someone who works outdoors will have more opportunities to move, but also more skin-cancer-causing sun exposure. Occupation is also tangled up with other major determinants of health, like income and educational attainment, which are examples of social determinants of health.[2] High-income households and communities enjoy better health, on average, because they can afford better food, health care, and so on.[3] These are modifiable determinants, but only to a degree. I can't choose to become a billionaire or inherit great wealth, but I did have a choice to attend university, which put me on a path to higher earnings.

But perhaps the most influential determinant of health is where you live. The places where you are born and raised, where you bring up your own children, or where you grow old have enormous bearing on whether you encounter malaria or survive a bout with cancer. Take Chad, a landlocked country in the Sahara region of North Africa. There, the life expectancy at birth is sixty and a staggering one in fifteen children dies in infancy.[4] Yet head northwest to Algeria, and life expectancy extends by nearly twenty years, to seventy-seven.[5] Just one in fifty infants in Algeria dies before their first birthday,[6] an infant mortality rate less than one third that of Chad's.

And even within a single country, place matters a great deal. In the United States, the nationwide life expectancy is seventy-nine.[7] But in Sioux County, North Dakota, home to the tribal lands of the Standing Rock Sioux Tribe, the life expectancy is sixty-eight, over a decade shorter.[8] Drive east nearly 2,000 miles to Suffolk County, Massachusetts, where the city of Boston and some of the world's best universities are located, and life expectancy stretches to eighty-one.[9] In other words, Suffolk County residents can expect around thirteen extra years of life compared to Sioux County residents.

What accounts for these differences? It's not just simple geography—longitude and latitude—of course. It's what comes together in a particular location. The infrastructure of a place—physical, social, and cultural—shapes so much of our lives. Some towns have clean, running water, while others still depend on surface water or untreated wells. Residents of high-income communities may have multiple teaching hospitals nearby, while many rural or developing areas have only small clinics run by volunteers. Some communities offer access to public services to get residents through hard times, while others have nothing of the sort. These are the most basic building blocks of good health.

Consider your own neighborhood. Suppose you want a new job, or even a new career. Are there schools in your area that provide relevant training? Are there employers nearby? How much do they pay? How does that compare to the cost of living, and what does that mean for your quality of life, and your ability to afford fresh food or prescription medication? What forces allowed you to choose that neighborhood over another that is less well-served? These elements—economic opportunities, educational access, income—all comprise the infrastructure of a place's public health landscape.

I often reflect on the power of place in the context of my twin pregnancy. For all that went wrong, so much also went right. I received care at the Johns Hopkins Hospital, widely regarded as one of the best in the world. The experts who cared for me are among a few dozen in the entire country with expertise in fetal surgery. As a Hopkins faculty member, I am sure I received privileged treatment. When I challenged the NICU staff about medical decisions or asked to speak with a senior member of our care team, my input carried weight. What, I wonder, would our experience have been if I had the same complications as a store attendant in rural Appalachia? My twin-to-twin transfusion syndrome (TTTS) may never have been caught, let alone treated so aggressively. The NICU may have been hours from my home, and I may not have been able to participate in my twins' care as fully. When I wake in the middle of the night from anxious dreams, I wonder if I would have been the mom of one daughter and two angels. Instead, all three are here with me as I write this.

And how did all that come to be? I lived close to Johns Hopkins, for one, and I am lucky to have good health insurance. But there are other determinants, too. The United States invests a great deal in medical and scientific research, which my twins and I surely benefited from, thanks in part to a wealthy tax base and political governance that prioritizes scientific supremacy. The National Institutes of Health, for example, funded the research that led to the TTTS treatment that saved my twins' lives.[10] And even setting aside my complicated pregnancy, White women like myself are three times less likely to die during pregnancy and delivery than Black women in my state.[11] Those are all determinants of health, each far upstream of how close my house is to Hopkins.

Zoom out a level, as you would on a digital map, and different factors become evident. The history, geography, and political context of a place come into play. Legacies of oppressive and discriminatory systems, such as colonization and enslavement, influence economic opportunity and self-determination.[12] In the United States, a host of racially motivated policies and practices relating to zoning, restrictive covenants, mortgages, and public housing (often collectively called "redlining") were implemented at the federal, state, and local levels over the course of several decades of the twentieth century. These policies restricted housing options for Black families (and, at times, those belonging to other racial and ethnic minority groups).[13] Black Americans were made to live in neighborhoods with materially poorer school systems, environmental health, and economic opportunities than those of their White counterparts, and these disparities persist today. Similar systematic marginalization has led to worse health outcomes for tribal communities from centuries of discriminatory practices.[14] The long shadow of other, similar historic injustices can be found around the world.

Place, and everything it influences about who we become and the things, tangible and intangible, we have access to, determines so much. When you look at it this way, as a web of determinants of health with place at the center, a great deal becomes clearer. Namely, public health is not merely a matter of distributing vaccines or tracing contacts. None of that is possible, or as impactful, without the infrastructure to make it real.

TAKE HAITI, one of the poorest countries in the Americas. In 2010, just one in six people in Haiti had access to sanitation facilities.[15] Four in five lived below the international poverty line.[16] A series of hurricanes and tropical storms in the early 2000s damaged infrastructure and sparked food shortages.[17] Political instability and violence followed, and many experts and aid groups warned that the country was fragile.[18]

Then, in 2010, catastrophe struck. The small Caribbean island suffered a massive earthquake that severely damaged what infrastructure the country had. The magnitude 7.0 quake struck just outside the nation's capital, Port-au-Prince. Entire city blocks were reduced to rubble, trapping residents inside. Roads were rendered impassable, cutting off access to parts of the island. Electricity and phone service were down for much of the country, preventing rescuers and residents from coordinating recovery. The air traffic control system at the island's largest airport was also disabled, which significantly delayed emergency aid workers from flying in from overseas.[19] The quake damaged or destroyed more than a quarter million homes, displacing 1.3 million people into temporary settlements, and 60 percent of hospitals were destroyed.[20] Even the Haitian government was temporarily hindered. Its headquarters were flattened and many senior officials were killed in the quake. Estimates of direct casualties stretched into the hundreds of thousands. Many more lives were lost from indirect causes, like lack of access to care for people with preexisting health conditions.[21] Aftershocks continued for days, deepening the devastation, and preventing emergency responders from rendering aid.

But the earthquake was only the beginning of a tragedy that stretched on for years. The events that followed sparked an epidemic that further devastated an island already shattered and brings us back to an old enemy discussed earlier—cholera.

The same year that Snow was tracking down the origin of cholera in London, Italian scientist Filippo Pacini deduced that cholera was spread not by toxic vapors, which was the prevailing theory at the time, but by a microscopic contaminant that could pass from person to person or through food and water.[22] The main concern with cholera is dehydration due to severe diarrhea. The treatment is simple.[23] For people who are able to drink fluids, an oral rehydration solution can be made from clean water with added salt and sugar to replace lost electrolytes. The solution can be made at home for pennies or prepared from ready-made packets. Patients who are too weak to drink can be rehydrated by a tube passed through the nose and into the stomach, or through intravenous fluids. With proper treatment, almost everyone—more than 99 percent of victims—survive.[24] All it takes is clean water, electrolytes, and a way to deliver the fluids. But without those basics, the mortality rate in severe cases can be as high as 50 percent.[25]

Prevention is also simple. Cholera epidemics are unheard-of in communities with adequate sanitation. The United States has not seen a major outbreak of cholera since 1866, thanks to near-universal access to clean water and sewage management.[26] Even when potable water is not readily available, boiling, filtering, or adding drops of chlorine solution will ensure water is safe to drink. And in communities without toilets, latrines dug at least 30 meters away from water sources will typically prevent waste from contaminating the drinking supply.[27] These simple steps can banish the "blue death" for good.[28] But in places like Haiti that do not have access to clean water, outbreaks are devastating and difficult to control. When effluent (human waste) enters natural waterways, environmental contamination can put millions of people at risk.

This happens more than you might think. For a disease that is preventable through access to clean water—a fundamental human right[29]—it remains appallingly common. Between one and four million people contract cholera each year, and tens of thousands die. Most outbreaks happen in the sub-Saharan region of Africa, and in South Asia, where the disease is thought to originate, but any community without proper infrastructure is at risk.[30] The number of people for whom that is the case reaches into the billions.[31] The most affected places are among the most vulnerable communities in the world: outbreaks in temporary settlements, refugee camps, and conflict zones are common.[32] The destruction of physical infrastructure in Haiti following the earthquake created precisely the kind of conditions where waterborne diseases like cholera flourish.

Globally, improved access to clean sources has whittled down the number of people who draw their water from untreated sources.[33] In Haiti though, where infrastructure was underdeveloped before the earthquake and all but nonexistent after the earthquake, clean water and adequate sanitation were inaccessible to most. This was true even for the aid-workers who flocked to the island to assist with the recovery. At least one of whom, we now know, was infected with cholera.

AT FIRST, it was a mystery how the outbreak began. The disease is not endemic in Haiti. In fact, the island had not seen an epidemic of cholera for nearly one hundred years.[34] Some experts speculated that the earthquake could have disturbed natural deposits of bacteria, perhaps releasing latent pathogens locked in soil into waterways.[35] But before long, rumors began to circulate. Haitian health officials and Associated Press journalist Jonathan Katz soon narrowed in on a darker truth.[36] The earthquake was not to blame, at least not directly. Humans were. A United Nations camp that housed foreign aid workers was draining effluent directly into the tributary system that webbed throughout the island.[37] As Katz wrote in one Associated Press dispatch, "A buried septic tank inside the fence was overflowing and the stench of excrement wafted in the air. Broken pipes jutting out from the back spewed liquid. One, positioned directly behind latrines, poured out a reeking black flow from frayed plastic pipe which dribbled down to the river where people were bathing."[38] Later, genetic samples of the bacteria in the outbreak were matched to those found in Nepal.[39] The analysis further confirmed inadequate sanitation at the United Nations camp likely sparked the outbreak.[40]

If Haitians had access to clean water, the camp's sewage leak would have been nothing more than an embarrassing episode. But what should have been an isolated case of cholera in a traveler exploded into one of the worst outbreaks in modern history.[41] In the first year, almost 200,000 Haitians were diagnosed with cholera, and at least

4,100 died.[42] The momentous toll was in addition to the hundreds of thousands who had perished in the magnitude 7.0 quake. The second year, over 350,000 more people became infected and nearly 3,000 died.[43] Year three, over 100,000 reported cases and nearly 1,000 deaths.[44] These numbers are surely underestimates, given the under-resourced medical and public health systems creaking under the weight of the combined earthquake and epidemic response.

Remember: all that is needed to stop cholera transmission is clean drinking water and basic waste management. Chlorine drops and a well-placed latrine will do in a pinch. To save someone infected from a preventable death, only clean water and rehydration salts are required. This infrastructure is so fundamental that most people in the developed world simply take it for granted. But not in Haiti. In Haiti, it took a decade for the cholera epidemic to be fully controlled.[45] That decade saw the preventable deaths of thousands of people who would never have become infected, never would have died of dehydration, had they lived in nearly any other country in the Americas.

HAITI BATTLED THE EPIDEMIC for years with little in the way of support. Public health professionals and aid organizations did what they could to stop transmission and provide care to people in need, but it wasn't enough. Crucially, it *could* never be enough. As long as the island lacked potable water and sanitation facilities, the risk of cholera would remain. The same structural weaknesses that made the island uniquely vulnerable to the epidemic also kept it from mounting a swift recovery. By the time the outbreak was finally controlled in 2020, nearly ten years after the outbreak began, it had grown to be the worst in recorded history with over 820,000 cases and nearly 10,000 deaths.[46]

. . .

As students of determinants of health know, the role of Haiti's impoverished *physical* infrastructure is only a portion of the story. Another portion is the *political* infrastructure. Imagine that cholera had been imported by aid workers in the wake of Hurricane Katrina, the devastating storm that flooded New Orleans in 2005. The political power of the United States would have demanded a conciliatory response—an apology, if not financial restitution. Haiti had no such power to leverage, and it was left without recourse. Consequently, it took six long years for the United Nations to admit that their facilities were responsible for leaking effluent into the river, despite robust evidence and enormous international pressure to do so.[47]

With the belated apology came a commitment of $400 million meant to help contain the outbreak and install proper water infrastructure.[48] But the same systemic inequities again intervened. The money that the UN pledged as reparation for its role in the epidemic never materialized. By 2021, only $21.8 million (5.5 percent) of the sum had been produced.[49] Of that sliver, less than half was spent by the end of 2020.[50]

Even the eventual end of the outbreak and the paltry restitution from the United Nations is not the last chapter in this saga. The deficient infrastructure that allowed cholera to take root in the months and years after the earthquake still exists. As of 2020, nearly one in four Haitians still lived without potable water.[51] Two thirds lived with only a rudimentary toilet, and nearly a fifth had no toilet facility at all.[52] Perhaps it's no surprise, then, that on October 2, 2022, two cases of cholera were diagnosed in the capital city of Port-au-Prince. The outbreak quickly spread, reaching nearly 77,000 suspected cases by December of 2023.[53] The cycle of panic and neglect rises and falls, and the perils of inadequate infrastructure are asserting themselves again.

IT IS TEMPTING TO CLOSE our eyes to these tragedies as occurring "somewhere else." In high-income countries, it's all too easy to distance ourselves from the public health problems of developing countries by thinking of ourselves as safe from their trials, and certainly not responsible for their problems. A closer look shows we do not have much to turn our noses up at. Even in the United States, there are neighborhoods where the simple act of drinking a glass of water from the tap is unwise. Lead pipes and service lines, relics of the first half of the twentieth century,[54] still deliver water to over 9 million homes.[55] The effects of this poisonous infrastructure are outrageous. As many as 500,000 young children have elevated levels of lead in their blood, which can cause irreversible brain damage.[56] Lead-poisoned children can suffer intellectual disability and behavior problems.[57] Their health issues derive from inadequate infrastructure, just the same as cholera in Haiti.

And it's not just lead poisoning. There are nearly 2,000 "Superfund" sites across the US,[58] a designation given to places with hazardous and extensive environmental contamination. Often, sites are former manufacturing facilities, landfills, and other industrial zones. Aberdeen Proving Ground, where I found the graves from 1918, was a former chemical weapons depot. The site still has contaminated soil, groundwater, and surface water despite thirty-four years (and counting) of remediation.[59] These, too, are matters of public health infrastructure and determinants of health.

But just as the lessons of superspreading can be turned to public health's advantage to improve contact tracing, the lessons of infrastructure can be leveraged, too. Just as deficiencies in infrastructure compound, the gains and opportunities do, too. There are plenty of other examples where investments in public health infrastructure not only have served their intended purpose but have returned myriad other benefits, too. Take the smallpox program. After eradication had been achieved, questions surfaced about what to do with the disease surveillance systems, vaccination teams, and expertise that had been powering the program. The answer came in the form of the Expanded Programme on Immunization, or EPI, which was established in 1974 by the World Health Organization to leverage the eradication program's infrastructure to bring routine vaccinations to more children.[60] Initially, the EPI aimed to protect children against six major diseases: measles, poliomyelitis, diphtheria, pertussis, tetanus, and tuberculosis.[61] It also sought to strengthen the cold chain system, which ensures the effective distribution of vaccines at the proper temperature.[62]

Over the years, the EPI has drastically reduced the global incidence of the diseases it targets, saving millions of lives and significantly improving the quality of life worldwide. The program's efforts were instrumental in declaring the Americas region polio-free in 1994, followed by the certification of the Western Pacific region in 2000, and Europe in 2002.[63] More recently, the program has expanded to include newer vaccines such as those against hepatitis B, Haemophilus influenzae type B, pneumococcal disease, rotavirus, human papillomavirus, and others.[64] The EPI also plays a role in coordinating global strategies against vaccine-preventable diseases, including monitoring disease outbreaks, collecting immunization data, and advocating for increased vaccine coverage. These achievements were all made possible by the public health infrastructure that comprised the smallpox eradication program.

Or take another, more recent example. The 2014 Ebola outbreak in West Africa was the largest outbreak of Ebola virus disease in history, primarily affecting Guinea, Sierra Leone, and Liberia. The outbreak began in December 2013 and continued for three long years, eventually tallying more than 28,000 cases and 11,000 deaths.[65] The true numbers may be even higher.[66] Through it all, public health experts worried that the virus would find its way to neighboring Nigeria, home to some of the largest, densest, and most internationally connected cities in the world. Those concerns soon proved prescient when a man in Liberia who had developed a fever after being exposed to Ebola departed from the hospital against the recommendations of his health-care providers and boarded a flight to Lagos, one of Africa's largest cities.[67] By the time he arrived in Nigeria, he was desperately ill. The man was admitted to a local hospital where he lied about his illness, leading to a misdiagnosis of malaria. But when the malaria treatment failed to help, his doctors suspected Ebola. The traveler was isolated, tested, and confirmed as the first recorded case of Ebola in Lagos. During this period, seventy-two people were exposed to the virus. Nigeria was now at a critical juncture, potentially on the cusp of an explosive outbreak.

But as with the Expanded Programme on Immunization, public health infrastructure built to manage one problem became foundational for managing others. Just two years earlier, Nigeria had established an Emergency Operations Center (EOC) as part of its polio eradication program. The EOC was staffed by experienced epidemiologists from the Nigeria Centre for Disease Control and Prevention, who acted swiftly to mount a response. The Ebola patient was isolated, and his contacts were traced and placed into quarantine. Sadly, nineteen people became infected. But the outbreak did not spread beyond those initial contacts, and Ebola in Nigeria was contained.[68]

Even the U.S. Centers for Disease Control and Prevention (CDC) has roots in historical infrastructure programs. The CDC was established in the 1940s as the Office of Malaria Control in War Areas, which was meant to combat mosquito-borne diseases in Southern states and Puerto Rico.[69] That small unit evolved into the Communicable Disease Center, which took on a broader mandate of managing infectious diseases. It's because of this history that the agency is located in Atlanta rather than Washington, DC, like most government agencies. Over time, the agency's mission has expanded further to cover a whole range of public health issues. Today, the CDC manages everything from infectious disease to environmental health to occupational safety, and more.

There is a quiet debate within the public health community over whether issues of infrastructure are public health problems. They are problems, certainly, but do they fit in the remit of public health? Skeptics fret that labeling everything a "public health problem" could detract from one of the field's core functions of disease control. From my perspective, the lesson from Haiti is clear: Inadequate infrastructure sets up a domino effect. Had modern sanitation infrastructure been common on the island, the cholera outbreak could not have grown so large. Infrastructure and determinants of health can make or break the path to public health progress. To ignore them, or cast them off as someone else's problem, is to give up the chance to do our best work.

5

TRUTH TELLING

H istorian Heather Cox Richardson likes to recite the aphorism: *history doesn't repeat itself, but it sure rhymes.*[1]

I learned about the 1918 influenza pandemic early in my education, thanks to a beautifully rendered book by John Barry.[2] From the comfortable distance of nearly a century, I read the scenes of the tragedy and elemental upheaval during that pandemic with interested detachment. Like mending homespun garments by candlelight or traversing cobblestone streets by horse, they were dispatches from a more primitive time. Then came the COVID-19 pandemic. The veil between the past and the present lifted, and stories of cloth masks and outdoor meals were elevated from dusty relics to a guidebook. Instructions for how to weather a plague.

Until recently, I harbored an unexamined assumption that the modern craft of epidemiology had evolved beyond the scrappy, ungainly attempts to slow influenza's spread a century earlier. If asked, I would have said I was confident that we knew how to navigate, if not control, a pandemic. But come 2020, that hubris was laid bare. The COVID-19 pandemic overran us, just as the influenza pandemic overran a world already brought low by the Great War.

Regrettably, one of the most persistent—and avoidable—missteps in our public health response was exactly what Barry warned against: "Those who occupy positions of authority," he writes, "must retain the public's trust. The way to do that is to distort nothing, to put the best face on nothing, to try to manipulate no one." To this day, public health struggles to live up to that teaching. The rhymes of history are not always the gentle, lulling sort. Sometimes they are piercing reminders that the failings and foibles of the past are somehow, still, our own.

It's not always lessons from our own history—public health history, in my case—but the echoes of others that confront us with a mirror. One such lesson that caught my attention is a curious story from the history of weather forecasting. For decades, it was official policy that the word *tornado* should not appear in any weather bulletins.[3] Readers of the telegraphed reports might find warnings of severe storms or high winds, but never *tornado*. In retrospect, the policy is obvious folly. But meteorologists at the time had their reasons, and those reasons are not so different from ones that I hear from epidemiologists today.

Storm forecasting was not advanced in the first half of the twentieth century, not like it is today. Meteorologists knew what sorts of weather conditions could produce tornadoes, but they were not able to pinpoint the time or the place. The inherent uncertainty meant that warnings would often come to nothing, a major liability for risk-averse scientists.

But it wasn't technical skill alone that kept tornadoes out of weather alerts. Officials feared that warnings of tornado weather would incite fear, even panic.[4] Probably, they mused, the cure was worse than the disease.

Having seen a tornado with my own eyes, it puzzles me that panic was deemed an inappropriate reaction. Fear seems justified. Helpful, even, if it motivates action. But somehow, that was the prevailing view until as late as the 1940s.[5] Officials felt that averting panic was more important than any duty to warn. As a consequence, many lives were lost.

In outbreaks of disease, things are not so different. Confusion and debate about whether, when, and what to tell the public abounds. In the early days of the COVID-19 pandemic, buckets of ink were spilled over the World Health Organization's fumbling attempts to talk of the "global outbreak," ostentatiously avoiding *pandemic*, the proper—and loaded—word for such conditions.[6]

Those and similar communication—or perhaps, more accurately, censoring—machinations have, almost without exception, flopped when the public came to understand what wasn't being said. Today's storm warnings have moved far beyond the days when "tornado" was taboo. What convinced weathercasters to speak more bluntly, and how can public health follow suit?

ON MARCH 18, 1925, the U.S. Weather Bureau issued a dreary forecast. Although winds blowing in from the south the day before brought unseasonable warmth to Illinois, with temperatures into the sixties, Wednesday was expected to be dark and gloomy, with rain likely.[7]

Four-year-old Lela Hartman paid little attention to the weather.

According to an interview she gave to her grandson some seventy-five years later, Lela and her family were headed to West Frankfort to visit her grandmother, who had been widowed a few years earlier.[8] Grandma Lipsey lived on a farm in the south-central part of Illinois, in the rural plains nestled between Missouri, Kentucky, and Indiana. The farm had a barn, an apple tree, and majestic oak trees—all good fun for young Lela. When they arrived, the Hartman family parked their practically new Model T in the farm's barn and set out to enjoy a warm day in the countryside.[9]

As the afternoon wore on, however, the fair weather began to turn. The sky darkened and the air grew heavy, just as the Weather Bureau had predicted. As the thick wall of clouds drew closer, Lela's parents grew wary and began to wonder about taking shelter in the farm's underground cellar. Lela's grandmother, though, was unmoved.[10] After a lifetime watching storms blow over the plains, it would take more than dark clouds to impress her. The rest of the family, reluctant to leave her side, watched with rising apprehension as the angry curtain of clouds drew closer and closer. Finally, the sky was "almost black as night."[11] To other observers in the storm's path, the clouds were gold, red, orange, and purple.[12] Beautiful—and ominous.

Eventually, the power of the incoming storm could not be denied, even by Grandma Lipsey. The family took refuge in the cellar, huddled under the thunderous roar that heralds a tornado's arrival.

After the sound from the storm died down, Lela's dad labored to free them from the cellar. Young and strong as he was, he could only barely pry open the door. A tree had fallen over the cellar's only exit, nearly trapping the family inside.

Once outside, they were shocked at the profound devastation they found. The family's Model T, originally snug in the farm's barn, was sucked up, spun around, and deposited near the property's fence.

Somehow, the top had been sheared off like a grotesque convertible. Moreover, the barn that had housed the Model T was missing altogether, and the old farmhouse was turned on its foundation. Two gigantic oak trees on the property were splintered as if they had been axed by a vengeful giant.

In the wreckage, the family found a lady's scarf that was prettily embroidered with a basket of flowers. It may have been blown in from miles away. Residents of the storm's path would eventually find items in the wreckage that had been carried more than fifty miles.[13] Lela's mom kept the scarf as a kind of memento. "My mom used that for many a year. It was always a reminder of where it came from."[14]

Communities across the region endured similar destruction. As the *St. Louis Post-Dispatch* described the storm the following day, "the air was filled with 10,000 things. Boards, poles, cans, garments, stoves, whole sides of the little frame homes, in some cases the houses themselves, were picked up and smashed to earth. And living beings, too. A baby was blown from its mother's arms. A cow, picked up by the wind, was hurled into the village restaurant."[15]

THE HARTMAN FAMILY HAD SURVIVED the Tri-State Tornado, the deadliest tornado in US history. The storm killed a record 695 people and injured over 2,000 more.[16] Its size was so immense that observers struggled to make sense of it, like an ant gazing upon an elephant.[17] Unlike most tornadoes, which appear as narrow pillars of tightly twisting clouds, the Tri-State Tornado was so wide—at times it spanned a mile—that onlookers perceived it as a wall of clouds. Entire towns were reduced to rubble in a matter of minutes, with no warning except the sight of the impending vortex. In Parrish, Illinois, 497 of 500 residents were reported injured or killed.[18] In De Soto,

Illinois, a grain elevator was carried 40 feet by the wind.[19] Murphysboro, Illinois, was razed, with sturdy brick buildings collapsed into ruins. So shocking was the devastation that when a passenger train arrived at one demolished town, the crew stopped, then reversed course three miles to the nearest town to fetch help.[20]

IN THE HEARTLAND OF AMERICA, tornadoes inspire awe and fear. Unlike hurricanes, which form offshore and can lumber across the water for days before reaching land, tornadoes strike with little warning. A tornado with 150-mile-an-hour winds can uproot large trees, flip train cars, and unpeel roofs from homes.[21] The wind speeds of the Tri-State Tornado are estimated to have reached twice that, rotating up to perhaps 300 miles per hour.[22] According to the National Weather Service, "In the worst situations, well-constructed walls fail or are even removed. Large debris objects (cars, larger sections of roofs) become airborne missiles causing further structural failures. Trees debarked. Tornado wind speeds 113 mph or greater, but in worse case situations 260 mph or greater."[23] Moreover, tornadoes can move with surprising speed. A middling tornado travels over land as fast as a car on town roads. Some have clocked in at highway speeds. Split-second decisions to seek shelter underground or in an interior room, as Lela and her family did at the last minute, can mean the difference between life and death.

I have seen these dramatic storms firsthand. I spent some of my childhood in southern Iowa, one of several Midwestern states in "Tornado Alley." One storm in the 1990s destroyed my aunt and uncle's home. Although their neighborhood had a storm shelter, the storm came on too fast for them to take refuge there. Instead, they piled into the bathtub with their young son, a mattress pulled over top of the tub to protect them from flying debris.

On another occasion, I was in the checkout line at the grocery store with my mother when the town's tornado sirens began to blare. We looked out the store's plate glass windows to see the telltale cone of green-gray cloud stretching from the sky toward the ground. The tornado was a good distance away. But if you can see a tornado, you might soon find yourself in its path. At the sound of the siren, a quick-thinking grocery employee ushered everyone into the store's walk-in meat locker. Tornadoes are not just wind, but moving battering rams that can shatter windows and uproot trees. Even seemingly solid objects can become projectile missiles in the grip of a tornado's vortex. The thick, insulated walls of the meat locker would provide some protection from the high winds and debris. We huddled in the cold, surrounded by packages of raw meat, waiting as Lela and her family had some sixty-five years earlier for the roaring sound that accompanies an imminent tornado strike.

Thankfully, the store was spared. That capriciousness is part of tornadoes' mystique. The storms can wreak havoc on one side of a street, wiping out an entire row of buildings, while leaving the other side unscathed. The near-miss is seared in my memory as a brush with the tremendous—and mercurial—forces that Mother Nature can wield. The incredible damage on Grandma Lipsey's farm is a testament to just how destructive a direct hit can be.

THE TRI-STATE TORNADO was noteworthy not only because of its devastating brutality but because it was, to some extent, predictable. By 1925, the few tornado experts in the country had understood for decades the deadly alchemy of weather conditions that could produce the storms. But because of absurd notions about how people would react to tornado warnings, the public was never alerted to the threats.

The first person credited with establishing the science of tornado prediction was John Park Finley, a tall and portly man born in 1854 to a farming family in southeastern Michigan.[24] After earning a university degree, Finley enlisted in the US Army Signal Corps at a time when much of the military's might went to post–Civil War Reconstruction in the South and aggressive campaigns against Native Americans in the West. In science and medicine, though, the army was one of the premier institutions for research and innovation.

When Finley joined up, the Army Signal Corps was not even two decades old. It was established during the Civil War to support military communications.[25] To do so, it employed flags, signal lamps, and telegraphs—all tools that were useful to observe and communicate weather alongside military matters. And back then, the Signal Corps was the only game in town. There was no National Weather Service or civilian equivalent responsible for the country's weather forecasting, even though there was already widespread appetite for weather information.[26]

Finley was a natural fit. He had already earned a degree in meteorological sciences, putting him well ahead in the nascent field. To advance his education further, corps leaders sent him to the Signal Service school in Fort Whipple (now Joint Base Myer-Henderson Hall), Virginia.[27] Soon after, he began studying tornadoes in earnest.

His commitment to his studies was intensified by an assignment to survey the devastation left by tornadoes in the Central Plains regions, where tornadoes are common. The experience left a deep mark on the young soldier. He dutifully recorded "agonies of death," including horrific dismemberment of women and children caught up in the storms.[28] Survivors, he saw, "were still covered with specks of fine dirt and leaves which seemed to be driven into the flesh" despite numerous washings.[29] The written records of the scenes are tense with sorrow and horror.

The devastation Finley witnessed fueled what became a lifelong fascination with tornadoes, one to which he "devoted all his spare private time."[30] Eventually, the entirety of his decades-long career was devoted to the pursuit. But mere study was not enough. Finley's storm assessments happened only after the fact, when the damage was already done. He felt that instead of limiting themselves to retrospective analyses, the Signal Corps should anticipate and even warn of impending danger, so that residents could take shelter and secure their property.

Finley told corps leaders as much, advising in 1881 that they should establish a summertime storm unit in tornado-prone Kansas City, Missouri.[31] He imagined that the unit would function by receiving weather bulletins from the Washington regional office, which would then be forwarded to local telegraph stations across the area at risk. Conveniently, telegraph communications were a specialty of the Signal Corps. It would be perfectly in their strengths to transmit weather observations across their networks.

Although Finley's idea seems obvious now, at the time it was innovative and, to some, disruptive. Then, as now, bureaucracies were loath to assume new responsibilities, especially if doing so meant drawing attention to themselves. Although Finley's early work earned him a promotion, his ideas were quietly shelved. In a congressional inquiry some years later, Cleveland Abbe, Finley's superior, admitted that the recommendations "mysteriously and suspiciously" went missing, and were never put into practice.[32] It took decades for a warning system like what Finley proposed to be implemented. In the interim, thousands of lives were lost to storms.

In 1884, a new assignment in the Fact Room at Fort Myer afforded Finley a second chance to advance the field.[33] The Fact Room was a small unit where research studies were conducted into the nature of weather. Finley used his time there to pore over nearly a hundred years of tornado events, spanning 1794 to 1881. He studied the meteorological conditions that produced six hundred tornadoes and the impacts they caused.

From this careful analysis, Finley penned a set of rules that he believed could describe the atmospheric conditions capable of producing tornadoes. There must be a well-defined low-pressure area, and high temperature gradients, he said, and wind velocities must be increasing in certain quadrants.[34] Taken together, the rules could serve as a sort of checklist for meteorologists to evaluate whether tornadoes were possible on a given day.

Finley was once again ahead of his time, and his recommendations quickly became bogged down by internal politics. His peers squabbled over the accuracy and importance of his rules—while Finley's own analysis showed that he could correctly judge if weather conditions could produce tornadoes over 95 percent of the time and with at least five hours of advance warning,[35] not everyone agreed. His detractors took issue with Finley's approach of lumping tornado occurrences and non occurrences into the same statistics.[36] Independent analyses showed that while Finley's rules were quite reliable in predicting that conditions would *not* support tornado development, accurately forecasting tornado occurrences was more chancy.[37] Like his proposal for a telegraphic warning system, his system for forecasting tornadoes was never put into meaningful use. Instead, people like the Hartmans were left with no means of foreseeing tornado risk, despite the fact that the science was available to give some degree of advance warning.

The resistance to Finley's work ran deeper than disagreements over how to tally performance. At the time, there were serious doubts about whether actively warning the public of severe storms had value at all. Leaders in the Signal Corps offered various mercurial reasons for their opposition.[38] In their justifications, I hear eerie echoes—or rhymes, as Heather Cox Richardson would have it—of many of the fallacies that public health communicators cling to today.

For instance, officials feared that there was too much randomness in tornado activity for warnings to be useful to the public. Cleveland Abbe, the man who told a congressional committee of Finley's "mysteriously" missing recommendations, registered a curious list of objections. Abbe supposed that because tornadoes are capricious in nature, it would be all too easy to issue warnings that failed to materialize. Forecasters, he said, had "no right to issue numerous erroneous alarms. The stoppage of business and the unnecessary fright would in its summation during a year be worse than the storms itself."[39] What's more, Abbe felt certain that telephone operators would prioritize their own safety over a duty to warn, and abandon post before passing on the message. Ergo, the Signal Corps needn't bother. These sets of considerations, of course, apply equally to disease outbreaks.

One local weather section director, J. I. Widmeyer, had concerns of his own. Against all good sense, he argued that taking refuge in a damp basement would be more hazardous to health than the slim possibility of a direct tornado hit. (Personally, I'd take my chances with the basement.) Widmeyer also registered concerns over the "nervous troubles" that warnings might incite.

And besides, Abbe reasoned, "the certainty of destruction is absolute when the tornado comes, then it follows inevitably that there is no material advantage to be derived from any, even the most perfect, system of forewarnings and attempts at protection."[40] In other words, tornado hits are so catastrophic that there is no use warning in any case. A direct hit would mean certain destruction and death either way.

Scandals in the Weather Bureau eventually forced Finley out of his position. Although he eventually obtained the rank of colonel, he never received the recognition that he deserved for his scientific advances, and they were put to no meaningful use during his lifetime. He died in 1943 at age eighty-nine.[41]

DOES THIS SOUND FAMILIAR? Public health officials routinely shape their messaging based on perceptions or fears about how the public will react emotionally to that information. Take the earliest days of the COVID-19 pandemic. In the first weeks after news from China brought word of a novel SARS-like coronavirus, then Secretary of Health and Human Services Alex Azar appeared on television screens around the country. He intoned the same message, day in and day out: "the immediate risk to the general American public is low."[42]

As he said these words to the camera, with the nation's top health officials at his back, news from China came in fits and starts. From my desk in Baltimore, I followed the dispatches closely in those early days, using internet translation services to read news from foreign health bureaus and newspapers. The picture I assembled from the constellation of scraps I could gather suggested that the risk to the United States was anything but low. At first, reported infections were limited to people who had visited a specific market where the outbreak was presumed to have begun. But before long, health-care workers and eventually people with no known exposures were turning up sick. Meanwhile, the Chinese government was recommending increasingly stringent personal protective equipment and restricting the movement of citizens in affected areas. SARS-CoV-2 was clearly on the move. Yet each day, when I turned on the television, I saw Secretary Azar issuing the same banal assessment: "the immediate risk to the general American public is low." The truth of the matter, as the history books now reflect, was soon explosively evident to the entire world.

The platitudes and misdirections accumulated from then on. Later that spring, after the virus became firmly entrenched, so many seriously ill patients were hospitalized that health-care workers all but lived in the hospital, to the point where some developed oozing sores on the bridge of their nose and cheeks, skin rubbed raw by long hours wearing tightly sealed masks. The overwhelming demand for personal protective equipment caused a run on supplies. Hospital storerooms were dangerously bare of the equipment that health-care workers needed to protect themselves. Newspapers ran front-page stories of nurses on COVID units wearing trash bags instead of disposable medical gowns.[43] Masks meant to be used once and then thrown away were pressed into service dozens of times.

Naturally, the public wondered whether they, too, should be wearing masks to protect themselves against the virus. But against the backdrop of dire shortages among medical personnel, the suggestion that the average American should begin to wear a mask seemed impossible, even irresponsible. Moreover, the scientific data available at the time was discouraging. Very few studies evaluated mask use outside of medical settings. Of those that did, some research showed modest benefit, and some found that masks conferred no protection at all.[44]

One option was for public health officials to lay this out clearly. A warm and trusted scientist could describe the need to preserve supplies for health-care workers and summarize the mixed evidence for universal masking, perhaps in a press conference with a generous question and answer session. Somehow, nothing of the sort took place. Instead, public health officials claimed that not only was mask use not recommended, it should be avoided. The reasons they drummed up bear an uncanny resemblance to Abbe's protests against Finley's impulse to warn: If told to wear masks, people would panic. Masks were unnecessary, they wouldn't help. In fact, masks would actually *increase* risk, because people would wear them incorrectly or would adjust them by touching their face. (It's not clear how breathing unfiltered air was meant to be better.)

These were odd and brittle assertions that did not hold up to the slightest scrutiny. In parts of Asia, mask use had been common since the 2003 SARS pandemic. Surgical masks are sold in pharmacies there and people with minor illnesses wear them as a courtesy to prevent further spread.[45] Moreover, masks are worn by doctors and nurses all over the world, both to protect themselves and to protect their patients from infection. And although high-quality respirators like N95 and KN95 masks provide far more protection, health-care providers had worn surgical and even cloth masks to prevent infections in low-resource settings for years.

Eventually, health officials did update their guidance to recommend that people wear cloth masks in public. But the lesson many people came away with was not about whether and when to wear masks, but that public health officials could not be trusted to tell the whole truth. That cloud of suspicion shadowed the pandemic response for years to come.

SOMETIMES CONFUSED messaging stems not from fear of panic, but fear of stigma. During the 2022 mpox (formerly called monkeypox) outbreak, United States health officials went to comic lengths to avoid naming sexual contact as the primary risk factor for transmission. The reluctance was such that CDC's website on the virus mentioned almost nothing about gay men or sexual contact, both of which were defining features of the first six months of the outbreak.

This unfortunate communication decision emerged from the long shadow of the AIDS pandemic, born out of legitimate fear of stigmatizing gay and bisexual men. For decades, men were reluctant to seek testing and treatment (when it became available) for HIV, for fear of experiencing stigma and discrimination and in expectation of receiving poor care. This experience taught public health officials that drawing attention to risk factors concentrated in marginalized groups can result in *worse* outcomes, and in response they adopted cautious sensibilities.

While concerns about attracting stigma were deserved, speaking in code did no favors to the people most at risk. Gay and bisexual men, who at the time constituted 98 percent of mpox cases, were left to puzzle through the outbreak on their own.[46] Absent high-quality communications from trusted public health authorities, getting the word out about the painful new infection was left to the whisper network until nonprofit organizations and media publications serving the gay community took up the task of filling the void—which they did, as one must, with clear and unflinching information about who was at risk and what sexual activities were most dangerous—just as activists and publications had in the 1980s.

It's an approach that public health organizations like the CDC should have used themselves, right from the start. Clarity is a core tenant of public health communications. But instead, they attempted to control how important public health information made listeners feel—and it came at the expense of epidemiology's core mission of controlling disease transmission.

. . .

As historian Marlene Bradford notes, "In spite of 4151 tornado deaths from 1920 to 1939, including 794 [from the Tri-State Tornado] alone, the Weather Bureau did nothing to try to reduce the loss of life from these natural disasters." In 1925, "the state of tornado forecasting and warnings was as nonexistent in 1940 as it had been in 1870."[47] Had Finley's warning system been in place, perhaps his rules could have been used to advise residents of the risk. But his contributions were ignored, no one sounded an alarm, and lives were lost. From 1884, when Finley laid down his rules to the Tri-State Tornado, to 1925 and beyond, the same excuses and intransigence that constrained implementation of Finley's vision remained largely unchallenged. Tornado forecasting was considered too underdeveloped, the behavior of tornadoes was too capricious, and the public's reaction too unpredictable to justify issuing warnings.

As in tornado forecasting, so too in public health. Here we are in the twenty-first century, and the echoes of those days still resound. I suspect that public health officials push out dubious messages not with the explicit goal of misleading people, but out of a misguided belief that a reassuring message should take priority over a truthful one. This emerges from the desire of public health and political officials to avoid inciting panic, just as Finley's detractors feared reactions to tornado warnings. Panic means difficult questions, congressional inquiries, a dip in the polls. Panic means rocking the boat. But in truth, panic is rare. The need for truthful, comprehensive information is perpetual.

In March 1948, the weather forecasting community finally got the push it needed to change its perspective. The heroes behind this pivot were unlikely, and somewhat unwilling, vanguards. Captain Robert C. Miller was newly stationed at the Tinker Air Force Base weather station in tornado-prone Oklahoma. Both Miller and his colleague on duty hailed from sunny Southern California, where storms are rarely more than fodder for idle conversation about faraway places. This shortcoming soon became hazardous.

On the evening of March 20, weather headquarters in Washington, DC, sent the Tinker team upper air analyses—including data that later proved to be wrong. Miller looked the reports over and interpreted them to mean that the base could expect gusty conditions, but an otherwise "dry and dull night."[48] He issued a forecast for high winds, but no storms. It came as quite a surprise, then, when a billowing thunderstorm began to blow in from the southwest. The meteorologists could see on their radar screen that the storm cell was moving quickly, and weather stations en route warned first of lightning and, before long, a tornado.

From his office at the Tinker station, Miller could see the twister approaching. According to an account he wrote in an unpublished book, it appeared as "visible in a vivid background of continuous lightning and accompanied by crashing thunder."[49] He watched incredulously as the tornado swept across the base, past the aircraft hangars where military aircraft were stored and toward the base's runway. So close was the storm that the window in the aircraft control tower shattered from intense air pressure, badly injuring the staff inside. Even Miller, the base's weather forecaster, was in danger. The window in the operations building where he was huddled blew out, filling the air with debris.

After the storm, base airmen surveyed the damage. Several hangars were destroyed, as well as seventeen C-54 Skymasters, two B-19 Superfortresses, fifteen P-47s, five L-4s, three C-45s, three C-47s, three AT-11s, one B-25, and one PQ-14.[50] Numerous buildings took damage to their roofs and windows, and "all overhead electrical lines between [buildings] 2121 and 3001 were destroyed, as was the 66,000-volt main feeder to the sub-station."[51] All told, the installation suffered over $10 million (in 1948 dollars) in damage.[52]

THE NEXT DAY, five generals descended on Tinker to review the damage and, Miller assumed, to end his military career. Miller and his commanding officer, Major Fawbush, waited their turn in front of the generals with the air of men going to the gallows.[53] As the installation's forecasting team, it was their responsibility to alert base leadership to severe weather. They had failed in their duty, and the consequences were likely to be severe. Now they awaited judgment—the options seemed grim at either negligence or incompetence—before a formidable panel of senior officers.

To their relief, and to the benefit of any American who has faced a severe storm since, the men were spared. The panel ruled that the storm's surprise appearance was not a personal failing, but an "act of God," unable to be forecasted in part because of the faulty upper air analyses from headquarters.[54] But the meteorologists were not without penalty. Major General Fred "Fritz" Borum, the commanding general of the Oklahoma region, ordered the men to take up study of the weather conditions that produce tornadoes in penance. Borum had spent much of his air force career as an aviator, during which time he developed a keen interest in weather. Now, as the man in charge of the Oklahoma City depot and Tinker Field, it was his intention that no such storm ambush would happen again.

The Tinker meteorologists took up the task with gusto. Indeed, Miller dryly noted that he became "most interested" in the topic overnight.[55] They immediately got to work studying the meteorological conditions that produce tornadoes, and in just a few days they developed a better sense of the patterns and observations that incubate the storms.

Borum's directive was not unlike the line of inquiry that had seized Finley's imagination some six decades before. It is fortuitous that Borum thought to put the men to the task. Since Finley's days, it remained the convention to avoid making tornado predictions public. The word *tornado* was still banned from public communications, and the justifications for this prohibition had not much changed from Finley's days. Leaders in the weather industry feared that tornado forecasts would do more harm than good, either because people would lose trust in the weather service if forecasts proved wrong, or from nebulous concerns about the "panic" the warnings would cause. As a result, forecasters might warn of storms, but never of tornadoes.

Within a week, that would change. This time, instead of being shelved, as Finley's had been, the work that Borum sent to Miller and Fawbush would change everything.

LESS THAN A WEEK after the tornado struck the Tinker Air Force Base, Miller and Fawbush found themselves facing a fear. As the pair studied the weather maps sent from the Washington office, they again saw atmospheric conditions similar to those that had generated the destructive tornado just days earlier. Their analysis, Miller later recalled, "resulted in the somewhat unsettling conclusion that central Oklahoma would be in the primary tornado threat area by late afternoon and early evening."[56]

Despite the knowledge wrought from their recent tornado study (to say nothing of their intimate brush with death), the two men were still reluctant to sound the alarm for fear of being wrong yet again. Wary of whether to proceed, they brought their suspicions to General

Borum, who harbored no such reservations. In just the few days since the tornado strike, Borum had already poured his considerable energy into establishing a new storm preparedness plan to protect the base.

"Are you planning to issue a tornado forecast for Tinker?" he asked.[57]

Miller and Fawbush ducked the question, hemming and hawing, knowing they would be making history as the first to publicly predict a tornado if they were to do so.

General Borum, a longtime military man, had no patience with their skittishness.

"If you really believe this situation is very similar to the one last week, it seems logical to issue a tornado forecast," said Borum.

The chances that two tornadoes would strike the same spot within a week were minuscule, the men pointed out. "Besides," they said, "no one has ever issued an operational tornado forecast."

"You are about to set a precedent," replied Borum.

And so they did. At 2:50 p.m. on March 25, 1948, Miller and Fawbush issued a tornado warning for the evening hour, the first such public warning ever issued. Base personnel leaped into action implementing Borum's new emergency plan. Aircraft were moved into hangars, incoming air traffic was rerouted to other airports, and base personnel, including those stationed in the control tower, were moved out of harm's way.

For their part, once the warning was issued, Miller and Fawbush had nothing to do but settle in for an uncomfortable evening. The storm was not due for several more hours, and the entire air force base—if not the entire Weather Bureau—was waiting to see if they got it right. The pair waited in trepidation, their anxiety mounting with each passing hour.

By 5:00 p.m., the outer edge of their warning window, nothing more than mild thunderstorms had passed through. Miller was distraught. First, he had failed to forecast a devastating storm. Then, just days later, he made the opposite mistake by sounding an alert to brace the base for a false alarm. He'd have to leave the military, he figured. Perhaps not even the local news station would have him. He imagined life as an elevator operator.

Despairing, Miller went home, leaving Fawbush to "go down with the vessel." Miller's wife, Beverly, offered some comfort, and he tried to go on with his evening—until the family's radio began to sound an emergency news alert.

"I was in another part of the house," he remembered, "but caught the words destructive tornado and Tinker Field."

"'Good grief,' I thought. 'They're still talking about last week's tornado—but why break into the news?'"

Then it hit him. The voice coming over the radio was alerting listeners to a new twister—the one that he and Fawbush had predicted.

Miller hopped in his car and sped to base. When he arrived, he was astonished by what he found. For the second time in under a week, the installation had suffered a direct hit. Power lines were down, and debris was strewn about. The vortex had also sucked up a B-29 aircraft left over from World War II. The losses stacked up $6 million more. But this time, because of their newfound forecasting skill and Borum's leadership, there was no loss of life.

More important, Miller and Fawbush had made history, despite some misgivings, as the architects of the first public tornado warning.

COMPARE AZAR'S FALTERING APPROACH to warning of the pandemic to come with that of another senior official. On February

26, 2020, the Trump administration was still insisting that Americans had nothing to fear, and the World Health Organization risked pulling a muscle from their vigorous dodging of questions about whether the tsunami of COVID-19 cases in parts of China, South Korea, and Italy constituted a pandemic. Epidemic watchers—those who could clearly read the signs—were in a liminal period. The gathering storm was obvious to those who raised their eyes to the horizon but unnoticed by most of the public and actively denied by elected leaders.

Against this static, a voice of clarity cut through. Dr. Nancy Messonnier, the head of the National Center for Immunization and Respiratory Diseases at the Centers for Disease Control and Prevention, participated in a press briefing on behalf of the agency that shocked the country out of its stupor.

"The fact that this virus has caused illness—including illness resulting in death, and sustained person-to-person spread is concerning," she said. "These factors meet two of the criteria of a pandemic. As community spread is detected in more and more countries, the world moves closer towards meeting the third criteria—worldwide spread of the new virus."[58]

Already, this frank assessment was a departure from the prevailing talking points at the time. But Messonnier had more to say.

"Ultimately, we expect we will see community spread in this country," she said. "It's not so much a question of if this [pandemic] will happen anymore but rather more a question of exactly when this will happen and how many people in this country will have severe illness."[59]

Contrast that with the remarks given by the Secretary of Health and Human Services just one day prior. The top-line message was, once again, that "the immediate risk to the general American public is low." Another senior official at the press conference reassured that the administration's "precautions are working" to contain the coronavirus and that the low number of confirmed cases was an "accomplishment."[60]

Confusingly, at other points in the conference, senior officials seemed to acknowledge that a pandemic was imminent. These incongruous messages did little to provide the clarity that Americans craved. Instead, they planted doubts in the minds of listeners that they were privy to the truth.

Messonnier did not fall prey to those foibles. The novel coronavirus was on the way, she said. And not only was its arrival inevitable, but it was also bound to be impactful.

"I understand this whole situation may seem overwhelming and that disruption to everyday life may be severe. But these are things that people need to start thinking about now. I had a conversation with my family over breakfast this morning and I told my children that while I didn't think that they were at risk right now, we as a family need to be preparing for significant disruption of our lives."[61]

BEYOND THE BREAK from reassurance, what I find remarkable is that Messonnier did not settle for the lofty talking points that filibuster airtime but do little to convey useful information. Talking about difficult subjects is, well, difficult.

When junior doctors are first confronted with the need to break bad news, many may be tempted to reach for euphemisms. A loved one has "passed on" they might say, or a prognosis is "poor." They adopt this approach out of a sense that indirect language is gentler, both for the giver and receiver. But some trainings discourage doctors from this tactic.[62] Better to be direct by saying a loved one has "died" or they are "unlikely to survive" a traumatic event. The reason is that in an intensely stressful situation, particularly one wildly unfamiliar like a hospital, it's very hard to absorb messages. Using plain language helps the listener to follow along.

As a patient, I've experienced this disorientation in myself. When my twins were in the NICU, sometimes I would leave a meeting in which I had nodded along the entire time, only to realize that I had very little idea of what was said—and I work in the medical field. Often, the facts would stick, but the context, or my ability to understand how to make sense of those facts, was absent.

The epidemiological equivalent of opaque language is fuzzy jargon like "domestic transmission" or "increase in cases." These phrases, which I have assuredly used, bear the same problems of euphemism. They are softer in the delivery, and they have the added benefit of accounting for uncertainty. What they don't do is give the listener any sense of the magnitude of the problem, or any direction about how concerned to be. By providing concrete examples of what disruption from the COVID-19 pandemic would look like to families, and how they might prepare, Messonnier gave them the tools to make sense of the public health "diagnosis."

"You should ask your children's school about their plans for school dismissals or school closures," she said. "Ask if there are plans for tele-school. I contacted my local school superintendent this morning with exactly those questions. You should think about what you would do for childcare if schools or day cares close."[63]

These communication choices are strong, and her warnings proved prescient. Within a few short weeks, schools in the US were forced to pivot to what became a protracted period of closure. In some districts, reopening was delayed for over a year. Apart from a few isolated incidents, epidemics had never before forced school closures in the US.[64] Despite her early warning, most schools had done little to prepare and were thus left scrambling to adjust to the historic, seismic shift to instructing and supporting students from afar. Perhaps more perspicuous communication could have encouraged better preparation.

The press conference was a moment of truth telling if there ever was one, and for doing so, Messonnier was effectively silenced. She soon disappeared from the public stage, and about a year later she left the agency and her decades of service behind. Still, the fight for truth telling went on, and partisanship continued to play a detrimental role in the pandemic response.

IN WEATHER FORECASTING, the biggest change from Finley's time to the era ushered in by Miller and Fawbush was not technical skill or methodology. The groundwork for forecasting had already been laid. The biggest change was in how meteorologists came to regard the role of the public in receiving and reacting to storm warnings. For many decades, experts assumed that warning of tornadoes would produce baseless panic. This assumption was never critically examined, or pressure-tested against reality. As we know now, that prejudice dramatically underestimates the capacity of the public to receive difficult messages and to make wise decisions based on that information.

The tornado warning issued by Miller and Fawbush, at the urging of General Borum, marked a turning point that permanently changed the responsibilities of forecasters by reorienting their attention toward people in harm's way. Since that fateful week in Oklahoma in 1948, tornado forecasting has become a cornerstone of the National Weather Service's (NWS) mission to protect life and property. NWS meteorologists issue some 1.5 million forecasts and 50,000 warnings per year.[65] In 2012, the NWS even took the unusual step of giving people almost a full day's notice of an upcoming period of high risk of severe weather.[66] The advance warning was credited with saving lives when a dramatic line of storms produced more than one hundred tornadoes in the span of a few hours across hundreds of miles in the Midwest.[67]

Today, the National Weather Service prioritizes communications efforts, including commissioning special studies of how best to communicate storm risk to the public even more effectively.[68] Which is to say, they continue to refine their approaches to reach more people more quickly, and to inspire more protective actions.[69] The meteorological community has even come to incorporate research and expertise in the social and behavioral sciences to improve what they say and how they say it. Newer initiatives take into account factors like how different vulnerable populations may understand risk communication, what cognitive biases may interfere with the public's understanding of storm warnings, and what framing is more effective at mobilizing people to take protective action.[70] In short, public warning has become central to the National Weather Service's mission. It has even gotten to the point that one might argue the pendulum has swung the other direction—with cable news and click-hungry internet sites overhyping weather events, perhaps inuring people to real risks.

And just as the weather industry evolved to prioritize messaging, so too should public health. The job of public health experts is to give people clear guidance about what they will face and how they can protect themselves. That does not include a special mandate to prevent panic or stigma when doing so sacrifices clarity. Although both objectives deserve a place on the priority list, they should rank well below the duty to warn and the duty to tell the complete and honest truth—no matter how difficult it is to tell and to hear. The impulse to offer unfounded reassurance or conceal tough truths is a dereliction of that duty. It's a mistake public health has been making for at least a century, if the failures of the response to the 1918 pandemic are any indication. It's a mistake we need not make again.

6

POLITICS

D r. Messonnier's sudden shift out of the spotlight struck a personal chord with me. Her conspicuous absence was interpreted by many (including me) as a clear indication of the escalating political pressures that were clouding an already complex health crisis. If a pandemic was on the horizon, as Messonnier cautioned and I feared, the emphasis from White House leadership should have been on preparing the nation. Instead, it seemed as though energies were directed toward minimizing the public's understanding of the severity of the situation. What we desperately needed in that moment was clear foresight and candid communication, not political jockeying to keep bad news out of the headlines. The political meddling raised unsettling questions about the autonomy of scientific institutions like the CDC. This not only chipped away at the public's trust in health agencies, but also sowed seeds of doubt about the administration's capacity to competently manage the impending crisis.

Tensions only rose from there. In the months that followed, a push and pull between public health priorities and political ones erupted. Public health experts, operating within their area of expertise, made recommendations singularly focused on fighting the pandemic. They suggested measures like stay-at-home orders and school closures with the goal of slowing transmission and preventing hospitals from becoming overrun. Yet when transitioning from guidance to governance, the waters became murkier.

Here is how one school of thought sees it: Most public health recommendations are advisory, meaning they are recommendations, not rules. The responsibility to enact measures like school closures usually (though not always) falls to elected leaders and school boards. And at the political level, there is more to consider beyond just health. Economic ramifications on businesses, disruption to children's education, and acceptability of pandemic control measures among the public are all relevant but beyond epidemiologists' scope. Health officials could suggest social distancing, but it is up to the politicians to weigh the broader implications—economic, social, psychological—and make the final call. When the "wrong" decision was made—a judgment that could only be made in retrospect, I might add—it was the public health profession that was unfairly blamed. The proper balance, proponents of this line of thinking would say, is for public health to give advice, and it's up to the politicos to decide what to do with it.

Over time, those messy, difficult decisions about pandemic-era decision-making fueled a powerful movement advocating for public health to stay far removed from politics.[1] The call has now echoed in opinion columns in national newspapers and has been discussed on cable news for years, solidifying it into dogma. The movement promotes a clear division between public health professionals and political leaders, similar to the FBI director's ten-year term, which is meant to ensure nonpartisan leadership.[2] The concept is now even reflected in formal policy decisions—President Biden's White House has established a high-level Scientific Integrity Task Force and directed agencies with a science portfolio to appoint a Chief Science Officer and Scientific Integrity Officials.[3]

I subscribe to a second, less popular school of thought: while I respect the sentiment behind separating politics from public health, I don't think it's so black and white. Leaving behind the political jockeying of Washington, DC, cannot and should not be a goal. I fear that in the current formulation, public health is trying to have it both ways. Health officials want to be able to communicate messages without political interference, but they don't always have insight or expertise into how those messages fit in with other considerations that inform decision-making. On the other hand, public health also wants to benefit from the support—both intellectual and financial—of policymakers and for them to express confidence in public health's recommendations.

I think it is wise to ask: If we succeed in "getting politics out of public health," what are we losing? It's not enough to merely present the data and hope for the best. To effectively protect communities and guide change, public health must step further into the political arena, not back away. Dancing around the edges of authority will only dilute the field's impact. It's time public health claims both the influence and the responsibility that these vital roles demand.

THE TUSSLE OVER Messonnier's truth telling was not the first time that political officials stepped on the toes of public health advice. Dr. William Foege, former CDC director, faced a similar dilemma. At six feet, seven inches, Foege possesses not just a formidable stature but a depth of experience that few can match. Early in his career he worked on the smallpox eradication campaign while on a missionary assignment in Nigeria. While there, he is credited with popularizing the surveillance-containment strategy that pushed eradication over the finish line when the original mass-vaccination approach began to falter. The assignment also came with considerable adventure. His wife and their three-year-old son were evacuated from the country when civil war broke out in 1967. Not long after, Foege returned to CDC headquarters, where he later went on to serve as the agency's director from 1977 to 1983.[4] That role afforded him plenty of brushes with politics, and not all of them were positive.[5]

One unsavory incident occurred in the late 1970s, when government researchers began to investigate troubling reports of an increase in Reye's syndrome, a serious neurological disease that devastated previously healthy children seemingly out of nowhere. Reye's syndrome victims were most often school-age kids who were recovering from common viral infections like influenza or chickenpox. Just when it seemed like they were on the mend, the children would abruptly deteriorate, sliding into lethargy and eventually a coma.[6] Some 30 to 40 percent of those stricken never recovered.[7] And even among the survivors, many grappled with permanent brain damage. For years, scientists were unable to find a cause for the rare illness, leaving parents afraid that their child could be next.

By the 1980s, the origin of Reye's syndrome began to come into focus. A series of small studies revealed a common thread.[8] Just as I give my daughters over-the-counter medicine when they aren't feeling well, parents of Reye's victims reported that they had offered salicylate, more commonly known as aspirin, to ease their child's symptoms. Careful epidemiological investigation suggested that the aspirin was triggering the rare neurological disease.[9]

Health authorities, including Foege, recognized the seriousness of the situation. By the end of 1980, at least 2,000 children had developed Reye's and hundreds had died.[10] If parents could be adequately warned of the danger, untold numbers of young lives could be saved. Public health leaders wanted to move quickly with a preventative measure: a warning label on aspirin bottles advising against its use in children with influenza or chicken pox infection.[11]

The aspirin industry, fearing a dip in sales, did not take kindly to this idea. The drugmakers launched a vigorous lobbying effort to fend off regulation requiring a warning label.[12] To do so, they devised a two-pronged strategy: first, they would manufacture doubt by downplaying the strength of the scientific evidence and seeking to discredit studies implicating aspirin as a risk factor for Reye's. At the same time, they would launch a nationwide awareness campaign of their own, proactively advising parents not to give aspirin to children with viral infections. This tactic was meant to rally public trust and weaken the need for regulation. Why *require* a warning label, the industry implicitly suggested, when the same message was already being shared voluntarily.

The political winds favored the drugmakers. The Reagan administration, in service to its commitment to deregulation, was reluctant to impose any restrictions on businesses. They pushed back on their scientists' recommendations, choosing instead to align with the aspirin industry by demanding that scientists produce additional evidence.[13] Stymied, the warning label rule was put on hold while yet another study was conducted.[14] When that, too, confirmed a link—and Congress began to apply additional pressure—the labeling requirement finally moved ahead. A product label was finally added to aspirin bottles in 1986,[15] five years after the link became apparent.[16] The cost of this delay was more than just time. One analysis suggested that the postponement caused 1,470 excess deaths—a wrenching reminder of the balance between commerce, politics, and public health.[17]

Foege was disgusted by the Reagan White House's interference. Years later, in a candid conversation with his alma mater, the University of Washington, he described his growing disillusionment. "Politics still got involved in public health," he said. "I did not stay much longer."[18]

Here we see clearly the costs of political interference. Political agendas sometimes have priorities at odds with public health, like short-term convenience or appeasing constituencies like the aspirin industry, rather than focusing on policies that will improve health. Those strategies might garner temporary support or skillfully squash a public relations crisis, but they betray a duty to protect and warn. Clearly, some degree of autonomy is necessary in matters of public health, safety, and security. Scientists must retain a voice, one that can be used to make frank assessments of situations within their expertise— even if those judgments are unpopular. Epidemiologists know epidemics, climate scientists know how our environment is changing, and toxicologists know when chemicals threaten our health. Those experts must be allowed to raise alarms, and when they do so, it's worth listening. But reserving the right to independent judgment does not mean that public health should abstain from politics altogether.

On the other hand, elected leaders like members of Congress, state legislatures, and governors have been elected to represent the public's interests. If they falter in that duty, they will be held accountable at the polls. Ultimately, it is they who have been granted the authority to make enormously complex and consequential decisions about matters of governance. Depending on their role, they control the power of the purse, and they hold the authority to make laws and regulations that can progress—or regress—public health priorities.

Consider the issue of money. Funding for government services like public health comes from two main sources: Congress and state governments. Both make decisions about how our tax dollars will be spent, including budget priorities and funding levels. It's no secret that public health programs often receive limited attention and resources. Disease surveillance programs, health education campaigns, maternal and child health programs: all run on a shoestring budget. Staffing, too, is severely constrained. Some public health departments must operate with minimal personnel, sometimes with only one or two workers serving an entire county.

It is tempting—and probably correct—to interpret this financial austerity as a lack of recognition of the value of public health. But not all government health programs are so badly cash-strapped. Take the National Institutes of Health. Although both NIH and CDC operate with the goal of finding ways to improve health, their coffers differ markedly. The NIH boasts an annual budget of around $50 billion,[19] which is over five times greater than CDC's.[20] And most years, that funding gap widens. In the complex and political arena of budget appropriations, the two organizations have fared very differently. How has NIH not just survived, but thrived?

Budget insiders tell me that part of NIH's success in securing funding is because it has taken a somewhat different tack than public health in making their case to political leaders. Rather than eschewing politics and all that it entails, NIH has embraced a more collaborative relationship with Congress—and they've been able to draw on a powerful network of stakeholders to help them do so. The largest direct beneficiaries of NIH funds are research institutions and universities, whose faculty rely on grants awarded by NIH to power their research programs.[21] And because some portion of each grant goes to overhead expenses like buildings and shared supplies, the institutions themselves benefit from those funds as well. Most universities retain a government affairs team to ensure that their interests—to include a robust continuation of grant funding—are represented on the Hill.

That horsepower is amplified by another major constituency of NIH: patient advocacy groups. The research that NIH funds is meant to advance care and treatment for the millions of people living with a medical condition, ranging from cancer to heart disease to Down syndrome to ALS. Each of those conditions, and the thousands more that fill medical textbooks, has a patient advocacy group. Like universities, these groups prioritize policymaking as a key aspect of their advocacy efforts. For each, I suspect that ensuring robust research funding is priority number one.

For example, when one of my family members was diagnosed with celiac disease, it was the NIH's health information site that I used to learn about the condition.[22] Later, I used NIH's clinical trials database to track candidate treatments for increasing gluten tolerance as they progressed through clinical trials.[23] Those resources are available in large part because groups like the Celiac Disease Foundation have made their members' needs known, both to NIH and Congress.

Public health, on the other hand, does not have as many powerful, committed advocates. This is despite the fact that the public is ostensibly a beneficiary of public health dollars, funding vaccines for children, air quality in public school buildings, and prenatal care for low-income mothers. One reason is that the biggest beneficiary of CDC funding is state and local public health departments, who are not permitted to engage with policymakers except when requested and approved to do so.[24] There is no equivalent to the American Heart Association for influenza, no Susan G. Komen foundation for cholera. There are no philanthropic foundations that represent the interests of families served by lead pipes—or if there are, they are relatively small. Even the physical presence of public health is in Atlanta, far from the halls of power. As former CDC director Dr. Julie Gerberding quipped in an interview, "So, if you want to play you got to be in the game and the game is not played in Atlanta—unless you're a fan of the baseball team there."[25] Gerberding's successor, Dr. Tom Frieden, estimated that he made over 250 trips from Atlanta to DC to meet with legislators during the eight years that he served as CDC director under President Obama.[26]

I agree with both leaders that distancing public health from politics, both literally and conceptually, only diminishes its reach. Rather than creating even more divisions, elected leaders should be recognized as partners—albeit ones with whom there are occasional disagreements. But to voluntarily erect a barrier, to cement public health's influence as advisory-only, risks missing out on opportunities to elevate health priorities and secure support for the big goals—like smallpox eradication—that create lasting, meaningful change.

One standout achievement in public health history illustrates the power of what can be achieved when politics and public health align. The program has proven to be an enormous triumph over a deadly infectious disease. The unlikely champion? Republican leadership, which defied traditional expectations by pledging billions in foreign aid to nations not typically at the top of our diplomatic list. The purpose? To combat a sexually transmitted infection on a global scale. And the real surprise? This initiative enjoyed unwavering bipartisan support for over two decades. How did this endeavor, swimming against so many currents, achieve such monumental success, and what does it teach us about the potential for collaboration between public health and politics?

IN THE EARLY 1980s, a hidden health crisis began to surface, first noticed in the busy hubs of San Francisco and New York. Young men, primarily from the gay community, were showing up at clinics and hospitals with rare cancers and infections that even experienced specialists had never seen firsthand.[27] The men were afflicted with a mysterious illness, initially named GRID for gay-related immune deficiency. As with the Legionnaires' outbreak in Philadelphia a few years earlier, the origin of the disease was elusive. At that time, no test existed to diagnose the illness. The earliest indicators were the opportunistic infections and rare cancers that preyed on the patients' failing immune systems, when little could be done. And by that point, in many cases, what we now know as the human immunodeficiency virus, or HIV, had already been transmitted to others, pulling them on to the course toward early death.

Within a few short years, the contours of the HIV epidemic began to shift. The virus extended beyond the sexual networks of gay and bisexual men. An increasing number of women became infected through heterosexual contact. And following the virus's discovery in the United States, it became clear that this was a worldwide issue, with epidemics sweeping across Asia, Latin America, and Europe. The crisis reached its zenith in sub-Saharan Africa, where HIV's toll was unparalleled.[28] The impact was singularly severe; unlike typical infectious diseases that predominantly affect the very young or old, HIV/AIDS ravaged populations in their prime, decimating communities at their most active and productive stages.[29]

By the mid-1990s, the advent of combination antiretroviral therapies (cARTs) began to turn the tide against HIV.[30] The potent drugs didn't just slow the virus, they converted a lethal diagnosis into a manageable, albeit lifelong, condition. People who once faced a certain death were suddenly, unexpectedly, able to resume normal lives. Moreover, as treatments continued to advance, patients on combination ART achieved such low viral levels that they could no longer transmit the virus.[31] The impact of ARTs was an extraordinary scientific leap and a vital rescue for countless lives, on par with the swift development of the COVID-19 vaccines.

But as with many new technologies, the miracle of ARTs was not shared equally among everyone in need. The drugs became widely available in high-income countries like the United States, but scarce in the parts of sub-Saharan Africa where transmission was most intense. Although aid programs worked to slow the epidemic and provide care to people living with the virus, meeting the full range of needs was an impossible task, and the public health community staggered under the size and scope of the problem. In the 1990s and early 2000s, life expectancy fell by twenty years in some countries in sub-Saharan Africa,[32] and millions of people lost their lives.[33] At one point, an estimated 37 percent of pregnant women in Botswana were infected with HIV.[34] Many of their children were orphaned by the same virus that would someday take their lives, too. As one observer mourned, "Not only were many households run by orphans, but entire villages were run by orphans, because everyone [else] was dead."[35] One estimate "projected that AIDS could be the worst epidemic since the bubonic plague of the Middle Ages."[36]

Senator Bill Frist, a practicing physician who became the Republican majority leader in 2003, recognized in the devastation a moral imperative. "History will judge whether a world led by America stood by and let transpire one of the greatest destructions of human life in recorded history—or performed one of its most heroic rescues," he said in remarks to his Senate colleagues. "President Bush has opened the door to that latter possibility."[37]

GIVEN THE HISTORICAL AND POLITICAL backdrop, President George W. Bush might seem an unlikely champion for AIDS relief. The Republican Party was often skeptical of expensive foreign aid programs, preferring instead to invest in domestic affairs. What's more, when the program was launched, the United States was still under the shadow of the dual terrorist and anthrax attacks of 2001. Around the same time, the president had authorized the controversial invasion of Iraq, opening a second front in the nation's ongoing war on terror. The national disquiet was further amplified by the SARS pandemic of 2003, the catastrophic disintegration of the space shuttle *Columbia* during re-entry, and a series of widespread blackouts that left many Americans without power.

Despite all these challenges, President Bush was determined to make the HIV pandemic a priority. In historical archives maintained by the George W. Bush Presidential Center, the president's inner circle described how plans to combat the pandemic came together.[38] The initial proposal put forward by HIV specialists from NIH was to invest in distributing a one-time treatment in several African countries that would prevent the virus from being passed from mother to child during pregnancy. The concept was ambitious—and expensive. The architects of the plan estimated it would cost $500 million, more than the National Institutes of Health's annual spend on HIV/AIDS research at the time.[39]

Mark Dybul, then a researcher at the National Institutes of Health, described the prevailing attitude at the time. "The world was moving, but moving slowly. People weren't thinking big. They were thinking a little here, a little there . . . no one was thinking billions . . . that we have to really get after this thing. We were thinking $500 million was a lot of money. But the President was thinking something much bigger."[40]

In early December 2002, the president's advisers convened in the Oval Office to discuss what more they could do beyond the maternal HIV effort. The group discussed whether it made sense to support the purchase and distribution of ARTs to partner countries. The medications promised to save lives, but patients would need to remain on the treatment for life. When asked for her feedback on whether to distribute ARTs, knowing they could help but not cure AIDS patients, National Security Advisor Condoleezza Rice replied, "Mr. President, one of the saddest days of my life was when my mother died. But I have always been grateful that she survived her initial bout with cancer. She died when I was thirty, not fifteen. That meant she saw me grow up, graduate from college, and become a professor at Stanford. You may not be able to cure those mothers in Africa, but maybe they'll live long enough to see their kids grow up. That will matter."[41]

At the end of the meeting, the president made a decision. He would announce it at the State of the Union address the following month.

Michael Gerson, the president's director of speechwriting, remembers the meeting this way. "I remember thinking in the aftermath of that meeting, that from a historical perspective . . . there had been other meetings around tables of government leaders in Moscow, and Beijing, and Berlin in the 20th century, where plans had been made to murder millions of people. And I got to be in a meeting where the President of the United States made the decision to save millions of people."[42]

. . .

PRESIDENT BUSH WENT TO CONGRESS and made a blockbuster request: $15 billion over five years to bring AIDS relief to fifteen countries in Africa and the Caribbean. The goals of the program were to prevent 7 million HIV infections, treat at least 2 million people, and provide humane care to 10 million people with AIDS. The program also proposed to provide care to children who had lost a parent or who were orphaned by AIDS. It became known as the US President's Emergency Plan for AIDS Relief, or PEPFAR.[43]

The legislation was not afforded an entirely smooth path on the Hill. Some conservative Republicans balked at the enormous expense and objected to program activities that they felt were immoral, such as delivering comprehensive sexual education instead of promoting abstinence and monogamy. Given the Republican majorities in both chambers, these protests could have been a significant roadblock. But President Bush did not waver. He personally lobbied members of his own party to support the bill and ultimately helped to secure its path. In the House of Representatives, the bill passed with an overwhelming majority, with a vote of 375–41. Passage in the Senate was unanimous.

Even that was not enough. "Some people call this a remarkable success," said President Bush at a World AIDS Day ceremony four years into the program. "I call this a good start."[44]

PEPFAR WAS A RESOUNDING SUCCESS from the beginning— so much so that the 2007 reauthorization of the program doubled the funding committed to $30 billion.[45] By 2018, "more than 14.6 million people were receiving lifesaving antiretroviral treatment, 95 million people had been tested for HIV, 2.4 million babies had been born HIV-free to infected mothers, and 6.8 million orphans, vulnerable children, and their caregivers had received support. The $80 billion investment has saved an estimated 17 million lives between 2003 and 2018."[46] The program has also trained tens of thousands of healthcare workers and supported the development of infrastructure and systems to improve access to HIV prevention, treatment, and care.[47]

Not all of PEPFAR's tactics garnered universal support. Particularly contentious was its emphasis on abstinence until marriage, a policy reflecting conservative Christian values, and a key aspect of PEPFAR's prevention strategy until 2008.[48] Critics, including Human Rights Watch and numerous health experts, actively challenged this focus. They contended that emphasizing abstinence proved ineffective in curbing HIV transmission and diverted resources from more effective educational initiatives, such as those advocating for correct and consistent condom use.[49]

Notwithstanding those debated aspects, the positive impacts of PEPFAR are evident and extensive. PEPFAR has supported the delivery of lifesaving antiretroviral therapy to millions of people, as well as scaling up prevention efforts such as condom distribution, HIV testing and counseling, and support for voluntary male circumcision.[50] The program is now credited with saving over 25 million lives and it permanently revolutionized the global response to HIV.[51]

THESE TRIUMPHS are the work of public health, and the people whose lives have been touched by the PEPFAR program are beneficiaries of what is unquestionably a public health program. But it is also the work of politics. It was political leaders—President Bush in particular—who made combating HIV/AIDS in Africa a leading priority. The United States Congress, another political body, has provided some $100 billion in investment.[52] To unlock small successes, it may be possible to operate without that kind of political support. But to make the sorts of large muscle movements that can set public health—and the world—on a new trajectory, there is no choice but to partner with political bodies. It is elected leaders who wield the power of the purse capable of dispensing millions or even billions of dollars to causes it deems worthy. Advocates and visionaries who can make a strong case for why their cause deserves support, and how taxpayers' precious dollars could be spent to effect change, will be in a much stronger position to influence change than those who think that the political realm is distinct.

That's not to say that inadequate advocacy is all that stands in the way of securing support for other social and health challenges. These are complex, multifactorial problems, and there are cadres of fierce and dedicated advocates rallying for many causes. But we cannot overlook that political support is a key critical element for them all.

I would like this for public health: Instead of focusing on how to get politics out of public health, we should learn from PEPFAR how to forge better partnerships that honor the contributions of each. Political interference can have both negative and positive effects. When gone wrong, it can prevent scientific messaging from reaching the public or hinder prudent crisis management, as in the case of the Reye's syndrome response. However, it can also unlock support for momentous innovations that have the power and reach to change the course of history. By embracing the complexities of political involvement, we can have a better shot at creating and sustaining lifesaving programs like PEPFAR.

7

COMMITMENTS

I t's often said that viruses know no borders. The original teaching of this saying is that individuals and governments should be concerned with what occurs beyond their borders and should therefore see some benefit to themselves in lending aid. But as I've spent more time in public health, I've come to understand a second meaning: a big part of the job is diplomacy. If viruses know no borders, then the choices and actions of our neighbors (which, given modern air travel, means nearly every community on Earth) can have profound effects on our public health at home. Had China been more transparent with the global community at the start of the 2003 SARS pandemic, for instance, the virus may not have gotten the head start that allowed it to reach more than two dozen countries. And had there not been universal support of pursuing smallpox eradication, it would never have been achieved. Ensuring everyone stays on course is essential for upholding good health everywhere.

This lesson has a home in the story of planetary protection, or the prevention of biological contamination, like microbes, during space exploration.[1] Planetary protection started as a bilateral commitment between the United States and the Soviet Union in the 1950s.[2] As the field evolved and more countries developed space programs, the number of individuals and organizations required to abide by the rules ballooned. Today, planetary protection has even taken on a commercial dimension as a growing cohort of space companies launch missions of their own. And as the number of spacefaring entities grows, so too do the stakes. The margin for error in protecting Earth from extraterrestrial germs (and vice versa) remains razor thin. Even a single lapse could have far-reaching and history-altering consequences.

In the lexicon of modern American ingenuity, the National Aeronautics and Space Administration, better known as NASA, has come to occupy a place of almost mythic reverence. The agency is synonymous not merely with rockets or astronauts but with the very idea of reaching beyond our earthly boundaries. When I think of NASA, I imagine astronauts tethered to space shuttles, enormous interstellar telescopes capturing impossibly distant galaxies, and the movement of planets in our solar system. Occasionally, the agency's work is even something of a spectator sport. Several years ago, my family and I made popcorn and settled in to watch a live broadcast. The ritual reminded me of tuning in to watch celebrities (another kind of star . . .) walk the red carpet before an awards show. But instead of appraising eye-wateringly expensive dresses, we streamed footage of an event taking place seven million miles from Earth.

NASA and its partner, the Johns Hopkins University Applied Physics Laboratory, were preparing to intentionally crash a spacecraft traveling approximately 14,000 miles per hour into an asteroid smaller than a cruise ship.[3] The maneuver was meant to knock the asteroid off its trajectory, like a goalie swiping aside a hockey puck. The deliberate collision was a test run of how the agency might protect our planet should an asteroid head our way. Miraculously, the gambit worked. The Double Asteroid Redirection Test (DART) impactor collided with Dimorphos at 7:14 p.m. EDT on September 26, 2022.[4] We cheered from the comfort of our couch when the spacecraft hit its mark. And just like that, the first planetary defense technology joined the national defense tool kit. These are the sorts of astonishing missions that NASA is known for.

But there is another side of NASA that is less well appreciated, one that draws on diplomacy more than science and technology. In 2017, a job posting for a role "concerned with the avoidance of organic-constituted and biological contamination in human and robotic space exploration" appeared online.[5] The posting noted a need for expertise in compliance, statutory requirements, committee work, and programmatic decisions. Rarely has a job ad seemed duller. But beneath the lifeless prose is work of considerable intrigue: space germs. NASA was searching for someone to fill the impressive-sounding role of Planetary Protection Officer. There is only one such officer at a time, and he or she is charged with overseeing the agency's mission of maintaining the boundaries between Earth's microbes and whatever pathogens may exist in the universe.[6]

It's a distressing thought, but sentient aliens are not all we have to worry about out there. Aliens could also be germs. Picture a valiant envoy from Earth, setting foot on a distant planet, navigating its unique terrain, thinking he hasn't encountered any life, only to return home with an unwelcome and unseen guest. It isn't difficult to imagine how a spacecraft could inadvertently become a breeding ground for exotic microbes, or how an astronaut could pick up pathogens during a spacewalk. Imagine that extraterrestrial pathogens—entities that our earthly immune systems have never encountered—found their way to our home. It's a chilling prospect. The potential ramifications, not just for humanity but for every form of life on our planet, could be profoundly catastrophic.

And it's not merely a matter of imagination. The transplant of pathogens from one environment to another has already touched off multiple disasters here on Earth. One of the most devastating on record occurred in the sixteenth century when Europeans introduced smallpox and other viruses' to the Americas. Prior to the arrival of the colonizers, the continent had been free of smallpox. As a result, there was no immunity in the Indigenous population, and therefore nothing to slow down the viruses' spread. It is estimated that perhaps 95 percent of Indigenous peoples may have died in the years following the invasion, most due to infectious diseases rather than armed conflict.[7] To put that in perspective, a modern event of similar magnitude would amount to the death of some 320 million Americans.[8] Fewer than 17 million people would remain, which is approximately the population of New York state.[9] Even today, concern about vulnerability to epidemic diseases is one reason why uncontacted tribes, like those that inhabit the Amazon region of Brazil, receive certain legal protections from the encroachment of outsiders.[10]

Going back further, a fourteenth-century epidemic known as the Black Death killed an estimated one third to one half of Europeans.[11] Parts of Asia and the Middle East suffered similarly devastating epidemics that decimated settlements. The cause was likely the bubonic plague, or *Yersinia pestis*, a bacterial infection that is carried by fleas that transfer the bacteria from rodents to humans. The name

Black Death comes from the blackened sores that appear on the skin of victims. So great was the devastation that some historians believe it led to a dramatic shift in the labor market.[12] In the fourteenth century, most local economies in Europe operated under the feudal system, whereby legions of serfs worked the land for a small number of wealthy landowners. But the balance tipped when so many people died from the Black Death, nobility and peasantry alike, that there were hardly enough survivors to carry on with the work. Some historians believe that landowners had little choice but to offer lower rents and better living conditions for the minority of workers who remained.[13] In some areas, the newfound power of the working class also spilled into the political arena, eventually eroding the feudal system altogether.[14]

Although *Y. pestis* is no longer the threat it once was, it does still cause outbreaks today. The disease is endemic in several parts of the world, including the island of Madagascar and the rainforests of the Democratic Republic of Congo. For instance, in 2017, Madagascar battled a large outbreak of *Y. pestis* that grew to over 2,400 confirmed cases, causing some 209 deaths.[15] Sporadic human cases are also diagnosed annually in the American West, usually in people who have handled animals.[16] Although modern medicine and epidemiology are now quite effective at controlling *Y. pestis* outbreaks, the infection remains deadly, even with antibiotic treatment.

While it still seems like the realm of science fiction, nobody can guarantee that an imported space germ would not proffer the same fate as those terrible plagues of history. And for that matter, fiction may actually offer valuable examples. Take Michael Crichton's 1969 novel *The Andromeda Strain*, which depicts a team of scientists battling a deadly alien pathogen, or the 1979 movie *Alien* and its sequels.

But perhaps the most eerie example is H. G. Wells's 1898 novel *The War of the Worlds*, which features a plot twist borrowed from the tragic lesson of epidemic smallpox in the Americas. When planetary protection officers consider the importance of their work, these are the kinds of worst-case scenarios that may come to mind.

THE GRAVE CONCERN isn't just about our world but also others. Forward contamination, the transfer of Earth's microbes to celestial bodies,[17] is one of the primary responsibilities that keeps planetary protection officers up at night. While it might not directly threaten humanity, forward contamination is more likely than pathogens imported to Earth and could have significant consequences.[18] Consider this scenario: After years of preparation and many millions of dollars, a mission to a distant planet discovers life. But there's a catch. How can scientists know it's an authentic discovery of a true alien life form? How can they be sure that it was not carried there by our own clumsiness? The best way to ensure that any discovery of life in space is a credible finding is to avoid bringing germs with us on our travels. Or consider a second scenario: If Earth germs were to make a home in space, they could become invasive, extinguishing or damaging the native life of the extraterrestrial body, just as earthly plagues have changed humanity's fate. Though the odds may be long, the consequences of botched planetary protection could be devastating.

NASA—and indeed the governments of every spacefaring country—have determined that gambling on microbial contamination is not to be chanced. Planetary protection officers are NASA's ambassadors, charged with guarding humanity and the as-yet-undiscovered life elsewhere in the universe.

. . .

THE IDEA THAT MICROBES must be managed during space exploration dates back to the very earliest days of the Space Race. The Cold War, marked by intensive rivalry between the Soviet Union and the United States, created a tense mix of geopolitical frisson and scientific competition. Throughout the 1950s and 1960s, the superpowers battled for dominance through their ambitious space programs as a proxy war for the two nuclear powers attempting to avoid direct conflict. Yet even while locked in competition, scientists on both sides of the globe recognized that any failures or misjudgments could result in catastrophe. The prospect of nuclear war underscored the fragility of human existence and the necessity for cautious progress in every scientific domain. In the case of planetary protection, such catastrophe could permanently devastate Earth, the moon, or humanity's ability to find extraterrestrial life.[19]

If there is one person responsible for codifying the role of planetary protection in space programs it is Joshua Lederberg, a scientific prodigy who graduated college at nineteen and won a Nobel Prize in Physiology or Medicine at age thirty-three[20] for his work on bacterial genomics.[21] His preeminence in microbiology soon grew beyond the confines of Earth. Around the time that he won the prestigious award, Lederberg became concerned by the biological implications of the Space Race. In 1958, shortly after the Soviets launched the Sputnik satellite, Lederberg pressured the National Academy of Sciences, the United States' premier nonprofit society of scientific professionals, to study and provide advice on precautions against breaking the barrier between humanity and any life that may live beyond Earth.[22]

In the early days of space exploration, the mission of Lederberg and his colleagues was straightforward: avoid contamination between Earth and other celestial bodies. Yet the specifics—defining what that entailed and devising the methodology—were anything but. Should a spacecraft be held to absolute, demonstrable sterility, or might minimal residual contamination suffice? This quandary, still debated among space afficionados, is more complex than it appears. Their recommendations would dictate an entirely new set of requirements for spacecraft design, engineering, and assembly.[23] Sterility guidelines would, in short, become a key hurdle in the race to space.

One proposal recommended "keep[ing] the probability of contaminating the solar system's planets (and in particular Mars) to no more than 1 in 1,000 during the anticipated period of biological exploration."[24] Supporters of that target felt that it was conservative enough to protect the planets that humans may someday visit while still being technically feasible for the engineers to achieve.

But not everyone was committed to the idea of maintaining even this level of stringency. There were scientists who felt that the United States was being too cautious in its planetary protection measures and that it came at the cost of progress. Skeptics feared that the efforts that Lederberg and his allies were proposing were reactions to a highly speculative threat and would come at a very real cost—the loss of the era's greatest scientific and political competition. The heart of this concern was captured in the letter from Norman Horowitz, an American member of a prominent group of self-organized scientists involved with early planetary protection. Horowitz wrote to Lederberg in 1960: "I would be willing to run the risk involved in a premature return trip, if a less bold schedule meant that a sample of Martian soil could not be brought back to earth in my lifetime."[25]

The tension between technical constraints on achieving sterility and the ambitions of the space program became a balancing act that continues to this day. Yet despite the rocky start, the objections of Horowitz and those who felt similarly were outmatched. The stakes of forever impeding the ability of scientists to find life on other planets—or to bring threatening life here—were simply too high. Even before the US's first mission to space, it became NASA policy to decontaminate spacecraft in the interest of planetary protection.[26]

Dissent aside, one factor in these early discussions was crucial: decisions were made early enough in the engineering process that they were built into space technology from the start. It doesn't always happen like this in technology development. Often, elements like safeguards or matters of human factors are relegated to an awkward retrofit. The early commitment spared Lederberg and his contemporaries the challenges of recruiting people who were already set in their ways and introducing new restrictions that would slow progress from an established pace. Although Lederberg and others did face objections from scientists and engineers who chafed at the added time and cost of sterilization procedures, those concerns surely would have been intensified if it were commonplace to proceed without them.

Despite the head start, the directive was immediately met with implementation challenges. NASA engineers tried and fell short each time in their attempts to sterilize the spacecraft in development. Engineers exposed equipment to gas mixtures. They irradiated it. They baked it. They tried chemical solutions.[27] All in an attempt to prevent germs from surviving on the hardware bound for space. Nothing they tried worked. The approaches teetered from being too gentle to kill pathogens to too harsh for the delicate equipment to tolerate.

In 1959, when the Space Race was in its infancy, both the skeptics and the proponents of planetary protection notched a bittersweet validation when a fumble reminded everyone involved that there was more at stake than being the first. Soviet scientists planned for the unmanned

Luna 2 to "become the first spacecraft to make contact with another celestial body."[28] The launch was a blow to the Americans' aspirations to be the first to establish supremacy as a spacefaring power.

But then, a twist. The Soviet spacecraft never made the return trip. On its descent to the moon, the *Luna 2* suffered a catastrophic failure. It crashed onto the surface, taking with it equipment that American scientists feared did not meet the planetary protection standards that had been agreed upon.[29] American scientists had no way of learning if their concerns were justified. The political conflict between the United States and the USSR was too fraught for scientists to exchange such sensitive details. The Americans feared that the moon may have already been irreversibly contaminated, and that if any signs of life were found on the moon in the future, suspicions would always linger that it was simply contamination from the *Luna*. To detractors of the United States' planetary protection program, the incident underscored the futility of sacrificing time and effort on the race to space in the name of planetary protection. If other space programs were not taking the same pains, what was the point of the US doing so?

OVER TIME, details emerged that reassured American scientists that the *Luna 2* spacecraft had, in fact, undergone some sterilization.[30] Moreover, it was eventually confirmed that the moon is barren of life and so Earth-originating microbes would have been of low consequence. Still, the near miss and its possible long-term consequences highlighted the need for a universal commitment to planetary protection from all those who venture to space. Any single mistake could lead to contamination of an extraterrestrial body, forever impeding scientific research.

Soon after the *Luna 2*'s launch, it was NASA's turn to test their mettle. The agency prepared to launch the first mission to land humans on the moon. The missions served as the first big test of both their aerospace prowess and their planetary protection efforts. Just as I sat on my couch to watch the DART impactor crash into the Dimorphos asteroid, some 650 million people tuned in to watch astronauts Neil Armstrong and Buzz Aldrin make history as the first humans on the moon.[31] As with previous American spacecraft, the outbound equipment was carefully decontaminated to minimize the chance that Earth germs would come along for the ride. But for this mission, there was an extra wrinkle: three human astronauts would be making the return trip as well. NASA scientists needed to protect Earth from them and any microbial passengers they might carry.

In what remains a point of national pride, the Apollo 11 mission accomplished what once seemed impossible: delivering three American astronauts—Neil Armstrong, Buzz Aldrin, and Michael Collins—to the moon. With Armstrong and Aldrin becoming the first humans ever to set foot on the lunar surface, this feat represented not just a significant technological and scientific milestone, but also a symbolic victory against the Soviets in the intense Space Race of the Cold War era. However, for the team in charge of planetary protection, the next phase in their critical work began upon the crew's return. Eight days after their historic journey commenced, the crew's re-entry vehicle pierced Earth's atmosphere and splashed down safely in the Pacific Ocean. Here, an intricate and meticulously planned procedure took place.

The journey home began with the US Navy's *Helicopter 66* plucking the crew out of the Pacific Ocean, miles below where the astronauts began their day. Before exiting the hatch, the astronauts and the helicopter personnel assisting them donned special biological isolation garments designed to keep space pathogens contained while the team was in transit. (Though the salt water that seeped into the suits raised doubts about whether the outfits were up to the task.) *Helicopter 66* then flew the group to the USS *Hornet*, a naval aircraft, where the outside of the biocontainment suits, the helicopter, and even the short path the astronauts walked on the deck of the ship were sprayed with disinfectant as added precautions. Once aboard the *Hornet*, the astronauts went directly to a specially designed Mobile Quarantine Facility, or MQF. To the delight of modern observers, the MQF was configured from an iconic Airstream trailer that NASA scientists had modified to maintain negative pressure so that air would not escape except when ventilated through a special filter.

President Nixon, on hand to welcome the astronauts home, was kept far away until the crew was safely ensconced in the MQF. A separate helicopter waited at the ready to evacuate him in the event of disaster.

Even that carefully choreographed sequence of events was not the end of the astronauts' journey. The *Hornet* navigated the MQF trailer, with the men inside, to Pearl Harbor, where the entire unit, fully occupied, was then loaded onto a military transport plane. The plane then ferried the group to their final destination: the Lunar Receiving Laboratory at the Johnson Space Center in Houston, Texas, where more comfortable quarantine accommodations awaited. The entire journey in the MQF took eighty-eight hours. The astronauts remained in quarantine at the Johnson Space Center for three weeks until it was presumed that any moon germs would have made their presence known. Finally, after enduring a helicopter, a ship, a trailer, a plane, and three weeks, the astronauts were given the all-clear.

The elaborate quarantine process was repeated twice more for the Apollo 12 and Apollo 14 missions. After each mission, NASA's team of scientists and microbiologists meticulously examined the returned spacecraft, the astronauts, and the lunar samples they brought back, looking for any signs of life. None were ever found. With this evidence of absence now well documented, NASA scaled back the requirements for missions returned from the moon, effectively retiring the MQFs.[32]

OVER TIME, the nature of space travel has changed. It is no longer powerful nation-states that control humanity's comings and goings. There are approximately one hundred commercial space companies working to develop and launch spacecraft—both with and without human passengers.[33] The growing cadre of commercial space organizations unaffiliated with any government have complicated efforts to ensure the planetary protection protocols are universally implemented.

Not all of these new ventures are committed to the principles that have kept our world (and others) safe. To wit, in 2018 billionaire entrepreneur Elon Musk made headlines by launching a sports car into space in a gaudy bid to prove the technical prowess of his spaceflight startup, SpaceX. The carcraft was piloted by a mannequin wearing a space suit, aptly named Starman. The vehicle (made by Tesla, another of Musk's companies) is now years into a destination-less journey through space, obliquely orbiting the sun. It may someday return to our corner of the universe, but until then, Starman is on a long and lonely journey through the cold night. Experts say that because Starman is not bound for a particular planet, the launch was not technically in violation of planetary protection requirements.[34] Still, the stunt was a flashy reminder of the changing landscape of space travel and planetary protection by extension.

For decades, the conduct of spacefaring nations has been governed by the UN's Outer Space Treaty, which includes as a principle that signatories "shall avoid harmful contamination of space and celestial bodies."[35] The treaty is supplemented by nonbinding policies set forth by the International Science Council's Committee on Space Research, to which all spacefaring countries have agreed to abide. Commercial spaceflight companies are bound by no such promises, only the rules of the nation in which they operate. No longer is space the exclusive domain of state-sponsored space programs. What this means for the next chapter of planetary protection is not yet clear. The space community is facing a set of challenges not unlike those raised by Lederberg and his contemporaries. The future of spaceflight is changing, and biological management is at risk of falling behind. A new generation of leaders will need to come together to chart a modern course for planetary protection on Earth and beyond.

8

SURPRISES

I f public health is a fabric, then the most visible threads, the ones
that consistently draw our attention, are those that have frayed. It
is the epidemics, the contaminated food supplies, the spike in infant
mortality that grabs our attention. And rightly so, for those threads
need tending. But what of the other threads, those woven tightly
against what would otherwise ail us? It may seem as if those holding
strong require less attention. But in fact, they trouble me, too.

The question I ask myself is this: How do we know that things are
going well for the right reasons? Do we owe our good fortune to luck
or skill? How can we be *sure* that what we are doing is working? We have
the outcome we want, sure. But with no prominent saves or near-
misses, it is impossible to say for sure what weaknesses we are missing.
How can we anticipate surprises—the unknown unknowns, as Donald Rumsfeld would say?[1]

If there is a canonical example of this in public health, it is the use of biological weapons. Imagine a scale of "things we are prepared to tolerate." Cholera is an affliction we allow to persist (apparently, defying all sense). Intentional outbreaks are not. The United States has spent eye-watering sums of money on preventing the development and use of biological weapons.[2] And the US is far from alone. In fact, almost every country in the world is party to the Biological Weapons Convention, an international treaty prohibiting the use of biological weapons.[3] This near-unanimity derives from shared agreement that the deliberate misuse of biological and toxin weapons is not merely immoral, it is "repugnant to the conscience of mankind."[4]

The high spending, the treaty, and the shared commitment to preventing the use of biological weapons has, by all accounts, worked. Prior to the ratification of the Biological Weapons Convention, interest in bioweapons was high, if the whispered secrets of defectors from the Soviet Union and North Korea are any indication. The now-dissolved USSR built an empire of biological weapons research during the Cold War.[5] North Korea's capabilities are likely to be somewhat more anemic but spine-chilling nonetheless.[6] And yet there are—mercifully—relatively few examples of deliberately caused outbreaks in global history. In the United States, there are exactly two. What can we learn from them, and what do they say about our ability to protect against future threats?

THE STRANGER of the two episodes began in 1981, when Bhagwan Shree Rajneesh fled his native Pune, India, for the freedom and prosperity of the United States. Rajneesh had developed a sizable following as a spiritual leader in India, where he preached a philosophy he called Zorba the Buddha, an eclectic mix of capitalism, sexual indulgence, and more traditional religious teachings.[7] Eventually, the good times in Pune waned. A combination of tax fraud and connections to drug-dealing and prostitution drove Rajneesh and his most devoted followers out of India and in search of a new place to settle, somewhere they could practice their devotions unencumbered. They found it in the form of a 64,000-acre plot of arid land in north-central Oregon, which they purchased for a tidy $6 million.[8] The land became the site of a cult scandal that shocked the nation.

Like any effective moneymaking operation fronting as a religious organization, the first order of business was to ensnare as many followers as possible. At this, Rajneesh excelled. He cultivated a steady stream of young people looking for free love, spiritual guidance, and family in all the wrong places. The cult's rapid expansion, fed by the tributaries of Rajneesh's opulent tastes and inflated ego, soon required extensive construction on the remote Oregon property. By the mid-1980s, what had begun as a scrubby plot of desert was transformed into something more akin to a college campus, complete with a sewer system, hospital, mall, police force, school, post office, and bus routes. Despite being zoned for agriculture, the compound grew to house an "overwhelming" number of inhabitants known as Rajneeshees.[9] They arrived from around the world to settle at the Rajneeshpuram compound and hear the teachings of the charismatic leader.

The influx of unmoored young people and their sometimes-salacious sexual practices provoked ire in residents of nearby Antelope, a tiny town of about one hundred.[10] There was little they could do against the bountiful coffers of the Rajneeshpuram. So, the people of Antelope turned to mundane bureaucratic labyrinths as tools of protest. It's a tactic that will be a familiar battleground for anyone caught in the clutches of a local school board or homeowners' association. Antelope residents levied zoning permits, county council seats, and over-policing for minor infractions as checks against Rajneesh's growing footprint.[11] Then in 1984, several years of administrative chess came to a head. A local election for county commission loomed, and with it hung the fate of Antelope and surrounding areas.[12] Two longtime incumbents who had previously run unopposed now faced Rajneeshee candidates vying for their seats. If the Rajneeshee representatives won, the tussles over whether the group could secure the permissions necessary to continue their expansion were all but extinguished. If they lost, the group's legitimacy and prospects of continued expansion would be lost along with it.

The Rajneeshees had already proven themselves unlikely to leave the vote to chance. Previous ballot measures that threatened their plans were defeated by their outsized constituency—but not all the votes were come by honestly. The group established a program they called "Share-a-Home" that brought in thousands of homeless people by bus from out of state.[13] The newcomers were given food and shelter, and in exchange they were encouraged to register to vote in local county elections—for Rajneeshee candidates, of course.

The upcoming election called for even more aggressive tactics. Ma Anand Sheela, one of Rajneesh's closest deputies, hatched a plot to tip the balance, this time not by stacking the vote but by suppressing voter turnout.[14] Sheela dispatched adherents to restaurants and grocery stores around the region with tiny vials of brown slurry, which they sprinkled onto grocery store produce and buffet-style meals at area restaurants. The revolting mixture was salmonella, a bacterial pathogen that causes food poisoning symptoms like vomiting and diarrhea. Sheela had purchased the bacteria through a medical supplier and incubated it on the Rajneeshee compound. Her plan was to incapacitate voters to deter them from turning out. Her plan more or less worked. An estimated 751 people fell ill, and at least 45 were hospitalized, making it the largest deliberate outbreak in US history.[15]

Between the efforts to inflate the rolls with imported voters and the mass poisoning, the plots were too bold to escape notice. FBI agents mounted an inquiry into the group's criminal activities, and what they found was chilling. Investigators discovered that Sheela and her coconspirators were willing to go much further than deliberate food poisoning in their bid for local control. The group had extensively researched poisons, chemicals, and bacteria in a laboratory they constructed at their compound. A later investigation revealed that they had considered using the bacteria that causes typhoid fever, but decided against it only because it might "attract too much attention."[16] Another idea, perhaps offered in jest, was to put dead beavers into the drinking water supply—potentially by liquefying them in a blender first—in hopes of spreading giardia. This plot was also discarded.[17] It eventually came to light that the group also plotted an assassination of a US district attorney and are suspected of having set a county court building on fire.

The people of Antelope were spared from further attacks when Rajneesh boarded a chartered flight to North Carolina, which federal investigators suspected was an attempt to flee from mounting suspicions of criminal wrongdoing. He was arrested when his plane landed, thus ending the era of Rajneeshpuram. The self-styled prophet ultimately took a federal plea deal and was forced to leave the United States. He died in India in 1990.[18] The compound was abandoned and fell into disrepair, and today serves as a Christian summer camp.[19] Sheela, for her part, served twenty-nine months of a twenty-year sentence for her crimes before being released on parole.

FOR DECADES, THIS SAGA WAS the lone example of the deliberate use of biological weapons in the United States. For public health officials tasked with biosecurity, such a rare and odd set of circumstances hardly seemed worth preparing for. But at the end of the twentieth century, Washington heard a steady drumbeat of warning from Dr. D. A. Henderson—the same public health luminary who had led the World Health Organization's successful program to eradicate smallpox. After leaving WHO, Henderson served in several senior roles in the United States government, earning a place as a heavyweight whose warnings could not be ignored. And so, when Henderson testified in front of a dais of senators and issued a stark warning at a 1999 congressional hearing, people took note.[20] He wrote in his testimony, "of the weapons of mass destruction, the biological ones are the most greatly feared but the country is least well prepared to deal with them." In the event of a biological attack, he continued, "[n]o one would know until days or weeks later that anyone had been infected." Henderson told the committee that responding to such an event would require extraordinary urgency, coordination, and resources, a prescient observation of the demands of the events to come—for which the United States was just as unprepared as Henderson warned.

Responsibility for countering biological weapons was (and often still is) grouped together with chemical, radiological, and nuclear threats. So-called CBRN (pronounced as individual letters or as seaburn) experts are charged with preparing for any and all the myriad ways that pathogens, toxins, chemicals, or radiation could be weaponized. But of course, each of those domains requires extensive, specialized knowledge, equipment, and planning. No single expert could reasonably stay ahead of it all. Moreover, to the extent that consideration was given to biological threats at all, CBRN expertise was largely confined to the military and law enforcement. Henderson felt that insufficient attention was paid to preparing the civilian world to recognize and respond to the use of biological weapons, a scenario he felt was uncomfortably likely.[21]

Many public health experts at the time were content to keep preparation for biological weapons at the bottom of the to-do list. The perpetually lean budgets that dictate much of public health's priorities were already stretched too thin. Adding responsibility for preparing for deliberately caused outbreaks—a threat that, at that point, had materialized singularly, in the form of the Rajneeshee food poisoning attack—would pull attention and programs from other, more pressing issues. After all, a common refrain in the public health community (then and now) chides that Mother Nature is the "ultimate bioterrorist."[22]

Henderson was not satisfied with this minimization. He feared that the risks and consequences of a deliberate biological attack were gravely underestimated. He saw, and sought to warn, that these gaps left the country vulnerable. In 1998, he established the Johns Hopkins Center for Civilian Biodefense Strategies at the school of public health where he served as dean for years (and where I have worked as faculty since 2017). It was as the center's inaugural director that he testified before the Senate that day in 1999, when he warned that the country was both vulnerable and ill-prepared for a deliberate biological attack.[23]

THESE WARNINGS soon rang in the ears of national leaders as they grappled with the events of 2001. If you ask any American what they remember about that year, you will almost certainly hear stories of the tragedies of September 11th. Nearly everyone can recount where they were and what they were doing when they heard the news of the attacks on the Twin Towers and the Pentagon. The contrast between the brilliant blue sky over New York City, the shocking scenes on television, and the fear they felt when shaken news anchors reported that multiple planes had been hijacked is difficult to forget.

The mood in the days that followed was grim, almost expectant. America was under attack and another strike seemed inevitable. Officials feared bioterrorism could be next and they hastened to prepare. *The New York Times* reported that Henderson was summoned to an emergency meeting with Secretary of Health and Human Services Tommy Thompson, where he was asked to give the secretary and his aides a crash course in dangerous pathogens.[24] As it turns out, their concern was prescient. But as is often the case in public health, what happened next still came as a surprise.

DR. STEPHEN OSTROFF, deputy director for infectious diseases at the CDC, was among those whose thoughts had turned to the threat of bioterrorism. Ostroff made a trip to New York City immediately after 9/11, expecting that city officials, already under immense strain from the terrorist attacks, would need extra help to keep watch. He brought with him a team of officers from the Epidemic Intelligence Service (EIS), a CDC program that trains clinicians and other health professionals in epidemiology. The team deployed to augment the city health department's search for unusual illnesses in emergency rooms, a signal that could reveal a bioterror attack. Ostroff got the team settled and returned to Atlanta.

Days later, on October 2, 2001, a tabloid photojournalist was hospitalized with a mysterious lung infection near Palm Beach, Florida. The patient, Robert Stevens, had recently returned from a visit to North Carolina when he fell ill with what his doctors believed was a case of meningitis, perhaps from an infection acquired while on his trip. Laboratory tests revealed something more troubling: Stevens was dying of inhalational anthrax.

ANTHRAX IS CAUSED by *Bacillus anthracis,* a bacterium resident in the soil of nearly every continent on the globe.[25] *B. anthracis* is known as an especially hardy bug. It is capable of forming spores that can survive for decades, waiting dormant until a chance encounter with a new host.[26] Humans rarely get infected by *B. anthracis.* When they do, it is often only after prolonged contact with infected animals or hides. Cutaneous anthrax, the most common form, is also known as woolsorter's disease because people who handle wool are at risk of developing the characteristic severe, necrotizing skin sores.[27] Without treatment, up to one fifth of people with cutaneous anthrax die.[28]

Gastrointestinal anthrax is the least common route of exposure. It develops after someone ingests undercooked or raw, infected meat— or, in a singular known case, after percussion of a hidebound drum wafted bacteria into the air that an attendee somehow swallowed.[29] And then there is inhalational anthrax, the most dangerous route of infection.[30] Also known as pulmonary anthrax, it develops when someone breathes in *B. anthracis.* The bacteria replicate first in the lungs and then move into the lymph nodes before launching an attack on the rest of the body. According to the CDC, "without treatment, this form is almost always fatal." Even with antibiotic treatment, around half of people who develop inhalational anthrax will die.[31]

Although *B. anthracis* does not spread from person to person, its propensity to form durable spores makes it difficult to control when it settles into human spaces. The spores are resistant to chemical, heat, and ultraviolet decontamination.[32] Fastidious protocols are required to kill the bacteria, and missing even a few spores can be an infection risk.

Together, the high mortality rate of inhalational anthrax and the challenges of decontamination make it fearsome. Its relative ubiquity in the environment renders it relatively accessible to would-be assailants. And although the natural formulation is a slurry that is not easily disseminated, it can be doctored into an aerosol formulation that can remain aloft and drift on breezes to contaminate wide areas.

The fear that the pathogen would be used as a biological weapon intensified when, in 1979, an unusual outbreak of dozens of cases of anthrax in Sverdlovsk, Soviet Union, was reported by a West German newspaper.[33] Word of the outbreak reached the ears of scientists in the West, who were immediately suspicious that the outbreak was not naturally occurring. Although the Soviet Union had signed on to the Biological Weapons Convention, the Cold War power was strongly suspected of maintaining an active bioweapons program.[34] However, Sverdlovsk was a closed city, and few people were allowed to come and go, so the source of the outbreak remained a mystery for some time. It was not until after the fall of the Soviet Union, when a top scientist defected to the United States, that details emerged of what went on in 1979. Apparently, the Soviet government was maintaining a secret bioweapons facility in the remote Siberian city. A breach occurred when a technician was dispatched to service an air filter that separated the facility from the outside air. He did not replace the filter, leaving the vent uncovered, but left a note for a later shift about the problem. The note went unnoticed, and facility operations resumed with the dangerous vulnerability. The next day, the error was discovered, and the filter replaced—but it was too late.[35] In the interim, clouds of aerosolized anthrax were released, settling over the facility and a neighboring ceramics plant. It was later revealed that at least sixty-eight people died, which is likely an underestimate.[36]

The one solace from the terrible event was that the winds were not blowing toward the main settlements in Sverdlovsk. If they had been, the death toll could have reached into the thousands.

BACK IN FLORIDA IN 2001, public officials reiterated that Stevens's infection could be nothing more than a stroke of bad luck. The day before his death, leaders at the Department of Health and Human Services said in a press conference that Stevens's infection was likely due to natural causes—but that they could not yet be sure.[37] President Bush said the case appeared to be an "isolated incident."[38] Tommy Thompson suggested that Stevens may have become infected while drinking from a contaminated stream, raising eyebrows among public health experts given that water is not a source of exposure.[39] But behind closed doors, public health officials felt uneasy. Most people with naturally acquired anthrax work closely with animals or their hides. Stevens had no such exposure. Moreover, wasn't an anthrax attack exactly what officials had been fearing after the September 11th attacks?

DECISIONS ABOUT WHAT TO TELL the public during an evolving public health crisis are perilous. Missteps are both easy and common. The Centers for Disease Control and Prevention's handbook on crisis and emergency risk communications advises officials to "be first, be right, be credible."[40] But more often than not, officials can hope for two of the three, at best. Being *first* depends upon beating journalists to the punch, a feat not easily accomplished when a big news story is brewing. Being *right* is also illusory. In a fast-moving crisis, what is right (or seems right) at the time of a press briefing can evolve the very next minute, and then again the minute after that. Worse, officials' interpretation of "what it all means" can evolve just as quickly, and the inconsistencies understandably rankle the public. And being *credible*, of course, depends upon a good track record of the previous two dictums. In my experience, it is the hardest attribute to secure and the easiest to lose.

AND SO IT WAS WITH Secretary Thompson's repeated reassurances that Stevens's infection was an isolated and naturally occurring incident. Just days after his remarks, two more employees at the media office where Stevens worked became infected with what test results revealed was anthrax.[41] Worse, environmental samples collected from the mail room at his work also turned up positive for anthrax. It was clear that something more sinister was afoot.

To public health officials, the death of Robert Stevens in Florida proved that fears of a follow-on biological attack were well founded. But it was yet another terrifying revelation that intensified the crisis. Another anthrax case had been found, not in Florida but in New York City. This time, the case was a staff member at NBC headquarters. It was discovered when epidemiologists were called to investigate a rash and suspicious sore on the NBC employee.[42] Part of the woman's duties, the epidemiologists learned, was to open Tom Brokaw's mail. On the victim's arm was a blackened sore—cutaneous anthrax.[43]

Ostroff immediately returned to New York City. But this trip, things were different. When he checked back into the hotel, he did so under an alias, acting on the advice of a colleague. Everything was about to change, for there was no longer any doubt. The anthrax infections were deliberate.

From the beginning, the response to the outbreak careened along two parallel tracks: the public health response to identify and treat cases and remediate buildings that had been contaminated, and the law enforcement response to identify the perpetrator.

On the public health side, in the weeks that followed Stevens's death, public health officials uncovered twenty-two cases of anthrax spanning not just New York and Florida but also Connecticut, New Jersey, and Washington, DC.[44] Most of the victims had been exposed to letters containing a suspicious white powder that sometimes puffed in the air when the envelope was opened. An additional 10,000 to 32,000[45] people were offered courses of antibiotics because they had likely been exposed.

Although the contaminated mail was addressed to prominent figures like journalist Tom Brokaw and legislators like Senator Tom Daschle, those letters never reached their intended targets.[46] It was the postal service workers, mail room attendants, and staffers who handled the letters who fell victim. Some developed cutaneous infections and others suffered more serious, inhalational anthrax. In a few instances, the victims were people with no obvious exposure or connections to people in high-profile roles.

Although Robert Stevens was the first case to be diagnosed, epidemiologists discovered that he was not the first to fall ill. Investigators uncovered infections dating as far back as September. The early cases, all of which were skin infections rather than the more severe pulmonary infections, were missed. The disease surveillance systems in place to detect unusual events had failed to sound the alarm until Stevens's death prompted a closer look.

The youngest victim was an infant who developed cutaneous anthrax after visiting the NBC offices where his mother worked. The oldest was Ottilie Lundgren,[47] a ninety-four-year-old Connecticut woman[48] who died from inhalational anthrax on November 21. Her infection was the last to be found, and the source of her exposure was never conclusively determined. In all, five people died from their infections, including Robert Stevens. The mailings dwindled and then stopped on their own within two months, just as mysteriously as they had begun.

On the law enforcement side, the Federal Bureau of Investigation and the United States Postal Inspection Service launched a criminal investigation that stretched on for years, ultimately accruing over 10,000 witness interviews and an estimated 600,000 hours of work.[49] Investigators determined that the letters were sent from a postal box in New Jersey, which is now on display at the National Postal Museum in Washington, DC.[50] Beyond that, there were few clues available to guide the case. At first, analysis of everything from the make of the envelopes to the handwriting on the letters came up empty.

The stuttering start was due in part because, in 2001, the field of microbial forensics was nascent. There were few specialized experts or scientific methods that could support the FBI's efforts to determine who had carried out the attacks. But over time, as the science evolved, new evidence yielded troubling clues. Although *B. anthracis* is found naturally in soil, scientists determined that the bacteria used in the attacks was not the kind that might have been harvested from the environment.

Bioweapons experts at the time suspected that if an attack were to occur, it would come from a terrorist group or as an act of war. But genomic analysis of the samples revealed that the anthrax in the contaminated letters originated from a single source—the anthrax stores maintained at the U.S. Army Medical Research Institute for Infectious

Diseases (USAMRIID), a military science facility in Maryland that conducts biodefense research.[51] Suddenly, the leading theory was that the perpetrator was not a foreign agent but an insider, someone who worked on biodefense themselves. Although deeply troubling, this discovery buoyed the investigation by dramatically limiting the number of possible assailants. Instead of considering every nation-state, terrorist group, and lone wolf anywhere in the world as suspects, investigators could focus on the few people who had access to the anthrax stores at USAMRIID.

Two suspects came into focus. The first, Steven Hatfill, was a biological weapons expert who had been involved in work describing how the postal system could be used to conduct an anthrax attack. Although the report was written with biodefense in mind, that report, coupled with Hatfill's expertise as a microbiologist and inconsistencies in the way he represented himself (he claimed to have a PhD that he never earned, for example), cast upon him a shadow of suspicion.

Investigators surveilled Hatfill for years, going so far as to tap his phone and raid his home. In one bizarre instance, investigators accidentally ran over Hatfill's foot and then ticketed him for "walking to create a hazard."[52] News media, too, dogged the scientist for years. Nicholas Kristof wrote a series of columns in *The New York Times* urging investigators to move more quickly against "Mr. Z," noting that if he "were an Arab national, he would have been imprisoned long ago."[53] *Vanity Fair* ran a lengthy piece by a "forensic linguist" who argued that the evidence lined up against Hatfill.

The inquiry into Hatfill dragged on for years, coming to a close only when investigators determined he did not have access to the anthrax stores at USAMRIID after all. Hatfill sued, eventually reaching a $5.8 million settlement with the United States government and additional settlements with media publications that had run articles accusing him.[54] (He resurfaced during the COVID-19 pandemic as an advocate for the use of hydroxychloroquine, a drug that became a popular folk remedy despite warnings from FDA against its use.)

Investigators next turned their attention to Bruce Ivins, another biodefense expert working at USAMRIID. Ivins had both access to the anthrax stores and the expertise to handle the pathogen. Moreover, in the course of the investigation, law enforcement officers learned that Ivins suffered from mental illness. He lost his security clearance after he was hospitalized for an acute psychiatric crisis. *The New York Times* reported that the hospitalization was touched off when, "With a strange smile, he told his therapy group that he expected to be charged with five murders and rambled on about killing himself and taking others with him, using his .22-caliber rifle, Glock handgun and bulletproof vest."[55] His therapist reported him to the FBI, which arranged for police officers to escort him out of his laboratory at US-AMRIID.

The Department of Justice announced Ivins was responsible for the mailings in August of 2008, seven years after the attacks.[56] But in one sense, it was too late. Ivins had died from an intentional overdose of painkillers a week earlier. The case was closed without ever going to court.

Following Ivins's death, rumors and allegations continued to swirl. Not everyone was convinced that the federal government had the right man. Skeptics allege that the evidence tying Ivins to the attacks was circumstantial, hinging on his access to *Bacillus anthracis* and the relevant technical equipment, plus his mental illness. To this day, the biosecurity community remains divided over whether Ivins was the perpetrator.[57] The details of who used anthrax to kill five people, and why, remain mysterious.

. . .

THE TRAGIC ATTACKS were shocking assaults on the homeland, and they revealed deep vulnerabilities in the nation's defenses— particularly because legislators themselves were targets of the anthrax-laden letters. For lawmakers in Congress, the work of fortifying the nation's biodefense systems could not begin soon enough. The compounding crisis of the 9/11 attacks and the contaminated letters spurred a bipartisanship and willingness to expend funds that is otherwise difficult to find in Washington. Congressional committees convened hearings and drafted legislation even while, as a result of the attacks, several congressional office buildings were closed for about three months while undergoing decontamination.[58]

The challenge, of course, was deciding what should be done. Although the attacks and the subsequent public health and criminal response revealed plenty of targets for reform, the mailings were just one scenario of many. Bioweapons experts had long warned that imagination is the only limitation for someone wishing to do harm using biology. What if the next attack came not in the form of anthrax, but smallpox, or tularemia, or even salmonella, as in the Rajneeshee attacks? Legislators were faced with the difficult task of sorting out what threats were most likely and which gaps were worth patching.

One man was ready with ideas. D. A. Henderson had been dedicated to preparing the country to respond to a biological attack for years, so it came as no surprise when he was tapped by Tommy Thompson in November 2001 to help chart the long path from vulnerability to preparedness. Among the priorities Henderson set during his tenure was the fortification of the nation's biodefense stockpile. A small holding of essential drugs and supplies was established in the years before the attacks; material from the stockpile was deployed after both 9/11 and the anthrax mailings to support the emergency response. Still, Henderson found the program to be anemic. He saw a need for a much larger stockpile with drugs, vaccines, and medical supplies that could be mobilized during an outbreak, whether it was natural or deliberate. Henderson had a particular interest in expanding the stores of smallpox vaccine in the stockpile as a hedge against the unnatural return of the virus he helped to eradicate.

He soon got his wish. In the aftermath of the 2001 attacks, a pivotal transformation unfolded. Within a year of the anthrax mailings, Congress ushered in the Public Health Security and Bioterrorism Preparedness and Response Act, a watershed moment that not only greenlit a substantial expansion of the national stockpile but also mandated defense against the specter of smallpox. The United States committed to maintaining a formidable armory of over 200 million vaccine doses, enough to vaccinate nearly every American.[59]

In the years that followed, the Strategic National Stockpile (SNS) became a vital resource in the nation's defense arsenal. Its remit extends beyond vaccines, encompassing the distribution of critical pharmaceuticals, personal protective equipment, and an array of indispensable materials. Looking back, the creation of the SNS is one of the most enduring and impactful reforms of that era. Over twenty years later, it is one of the most frequently exercised and effective resources in safeguarding the nation's public health.

The stockpile has not always performed flawlessly, of course. Some experts criticized the inability of the SNS to provide health-care workers with an adequate supply of high-quality masks and other protective equipment during the early days of the COVID-19 pandemic.[60] The SNS also ran short of ventilators, leaving clinicians on the front lines to weigh wrenching choices about who, among the sickest patients, should have access. Despite these shortcomings, never does the value of its existence come into question. The debate is always about how the program can do even more, never about whether it has value. The SNS has been called upon time and again to deliver aid in moments of crisis, and it delivers on its mandate.

Not all programs established in the wake of the anthrax attacks can claim such success.

In 2003, Congress allocated funding to the Department of Homeland Security for BioWatch, a discreet installation of air sampling systems in key locations nationwide in venues that could potentially be targeted in a bioterrorist attack. These air-monitoring systems use collectors to draw in air through filters, which are subsequently analyzed in laboratories to detect the presence of harmful pathogens. Defense experts hoped that in the event of an attack, the BioWatch Program would provide an early alert that something was amiss, giving authorities a head start in mounting a response.

Two decades later, BioWatch's future is uncertain. Over this period, it has weathered a series of critical reviews from the Government Accountability Office (GAO), the government's chief oversight body. The central critique revolves around the program's wan effectiveness, with a prevailing sentiment that, bluntly put, BioWatch doesn't live up to its intended purpose.[61]

One of the key points of contention stems from doubts surrounding its ability to effectively detect pathogens. BioWatch claims that it can detect attacks that could potentially result in 10,000 casualties. However, this assertion has met with skepticism from the GAO, which reported that the available scientific evidence does not substantiate the claim.[62] Pinpointing the exact threshold at which the devices can reliably detect pathogens is a tough scientific challenge, and source of considerable disagreement. Compounding these doubts is the perspective of many biosecurity experts, including myself, who question the utility of an air sampling system that can only detect events on an immensely large scale. Smaller, covert, and potentially more insidious threats may pose as much of a risk and are much more likely than massive attacks.[63]

And yet rarely does BioWatch come under serious threat of being defunded. Although its practicality is in doubt, and its capabilities are unlikely both in concept and execution to prevent or detect a bioterror attack, it has remained in place for years. As skeptical as GAO and some experts may be that it can achieve its objectives, there is also no firm evidence that it *can't*. It is the Schrödinger's cat of biosecurity: neither effective nor ineffective until it's needed, whereupon it will either work or it will fail.[64]

There is another force that keeps BioWatch in play. If the program were shuttered, it is unlikely that the funding would be put to better use, from a public health perspective. The temptation would be strong for lawmakers to move the money to combat another, nonbiological threat, or to shrink the budget in the name of fiscal conservatism. This puts public health advocates in a tight spot. Is something better than nothing? Should BioWatch continue if the alternative future has no biothreat surveillance at all, or if it instead buys a fraction of an F-22 fighter jet? The temptation is to let the program's weaknesses slide, or even put in a tepid good word. Such are the dilemmas that chronic underfunding in public health can produce. Public health proponents are so accustomed to making do with what we have that we sometimes forget to push for something better.

Compounding the uncertainty is the fact that no bioweapons attacks have been carried out since 2001. This state of affairs, of course, is ideal. The highest goal in biosecurity and public health overall is for nothing to happen, and to that end we have found over twenty years of success. But the quiescence does raise a delicate question: How do we know that what we are doing is working? We have the outcome we want, but with no prominent saves or near misses, how are we to assess where measures fall short, and what should we be doing to improve our preparedness?

My fear, of course, is that the work we have done to guard against threats will miss the mark, and the next big risk will come catch us off guard. It's these surprises-in-wait that weigh on me. Should that occur, it would not be the first time that public health was caught on its back foot. Whether it's the Rajneeshee food poisoning attacks or the anthrax mailings, odd turns of events seem to be the norm. And it's not just deliberately caused outbreaks. Most of the major epidemics of this century took forms that were slightly off center from what epidemiologists expected. In 2014, for example, Ebola virus disease, normally found in south-central Africa, defied expectations by appearing on a different part of the continent where it again surprised experts by spreading far more readily in dense urban areas than in the isolated communities where it had previously circulated. Then in 2015, a new virus that causes birth defects in infants born to infected mothers emerged as a mosquito-borne disease—neither high on lists of greatest threats.[65] And finally, the pandemic that overtook the world's defenses in 2020 was not influenza as nearly everyone had expected, but a coronavirus. While coronaviruses were known to have pandemic potential, most epidemiologists would have put their money on flu.

The sting of those surprises, and fear of what surprise could be next, is the disquiet that always lingers in the minds of epidemiologists. With these lessons in mind, we—both the public, and public health professionals—have no choice but to expect the unexpected.

9

TECHNOLOGIES

I am a member of the first generation of so-called digital natives, a designation that now seems quaint but serves as a reminder of just how recently personal computers were adopted into our every waking moment. Growing up, I chatted with friends online, played games online, and generally spent an unprecedented degree of my leisure time online. This did not sit well with my mother, who encouraged me to develop hobbies and friendships in the "real world." In other words, to follow a model closer to the one that every generation before mine has lived.

She was right, of course. Scientists now recognize that the near-complete overhaul of the way that teenagers learn and grow while "plugged in" to their devices has significant, negative consequences on adolescent development.[1] The mental health of young people has declined markedly over the last decade, a trend that many experts attribute to constant connectedness and the shallow intimacy afforded by those online relationships.[2]

But as a teenager, I would not hear of any concerns. Already the internet was changing the way people worked, played, and connected, and I saw no sense in clinging to the old ways. The irony is not lost on me when I now urge my own children to moderate their use of technology, particularly when the COVID-19 pandemic moved public education to Zoom. I became the one lamenting a life lived online. No longer did learning take place at a battered desk in a cinderblock classroom, as it had for generations. Pandemic kindergarten meant logging onto a laptop populated with a mosaic of black boxes with small children bobbing in and out. By the nine-month mark, having been out of the school building longer than she was ever in it, my oldest child was asking questions about dimly remembered notions like recess and the cafeteria. She noticed before I did that our lives were forever changed.

After the pandemic settled into familiarity, it became clear that we won't be returning to the way things were. Our school district recently announced that during snowstorms and other inclement weather, class will continue virtually instead of being canceled. That too represents a truism: once a new technology becomes entrenched, it is a permanent fixture until it is also replaced by something else in this constant march we call progress. This is the leading genesis of discomfort around innovation. Once adopted, there's no going back.

IN PUBLIC HEALTH, we face the same polarized tensions around the development and dissemination of innovation as any other field. New health technologies are introduced constantly, and often they are breakthroughs that change our health for the better. The bifurcated needle allowed smallpox eradicators to conduct efficient mass vaccination campaigns using just a fraction of the vaccine product, alleviating shortages, and reducing the need to train vaccinators. At-home diagnostic testing is putting more health decision-making in patients' homes, improving convenience and affordability. As we witnessed during the COVID-19 pandemic, vaccine platform technologies like mRNA promise to shorten the time to when a new vaccine can be designed and deployed. All are boons for infectious disease control.

On the other hand, there are drawbacks to any change, and those costs are not always given a full accounting. The pivot to online learning may have kept students and staff safe during the pandemic, but it also slowed academic growth and distanced children from the social supports and services that school provides. The cancellation of snow days—a cherished childhood tradition—is a more intangible loss, but one I mourn, nonetheless. The shift from laboratory-based diagnostic testing to at-home tests, which are not reported to public health authorities, has weakened our ability to track these diseases. And based on the performance of the COVID-19 vaccines, the protection that mRNA vaccines confer may fade more quickly than conventional vaccine types. These are downsides that can make innovations feel like a false promise.

In an ideal world, society would approach the introduction of new technologies much as a surgeon might approach a novel procedure: with deliberation, careful consideration of the risks and benefits, and, above all, with the intent to do no harm. But the real world is much messier. What is the right way—or even a way at all?—to evaluate whether new tech is for the best? Should we borrow a page from the playbook of presidential campaigns and introduce spirited debates and town halls to grapple with the implications of technological leaps? The Amish and Mennonites offer a thought-provoking model: they judiciously appraise technologies against their age-old values, often choosing to forgo technologies that would pull them away from their traditions.[3] Do we need a process like that?

The problem, of course, is that the landscape of risks and benefits is hard to define, and often very personal. It's akin to the complexities we were faced with during the pandemic, as educators, parents, and students weighed the trade-offs between Zoom classrooms and potential viral exposure. Experts are still undecided on which was more harmful, on balance. The repercussions of innovations can be multifaceted, long-lasting, and difficult to foresee. Predicting the many and various consequences of a technology before it becomes widespread, with sufficient lead time and clarity to influence its adoption, is a tall order. Yet, somehow, public health must do just that.

ONE SMALL ISLAND CHAIN off the southern tip of Florida has been at the forefront of the debate around the promise and peril of one new frontier in public health technology for over a decade. It also serves as an example of how an engaged community can make their voice heard.

The Florida Keys are home to a small number of permanent residents, a huge number of tourists, and an even bigger number of mosquitoes. It was the mosquitoes that led the agenda at a March 2012 meeting convened by the Florida Keys Mosquito Control District. Around four dozen local residents piled into a windowless conference room to hear a pair of presentations better suited to a graduate school auditorium than a Thursday evening town hall.[4]

Michael Doyle, the executive director of the Mosquito Control District, opened the session with an entomology lesson. There are nearly fifty species of mosquitoes in the area, he explained.[5] Most are mere nuisances, like the black salt marsh mosquitoes that thrive in the nearby swamplands. But two species threaten to make the islands all but unlivable, despite representing around 5 percent of the total mosquito population. The first is *Culex quinquefasciatus*, also known as the southern house mosquito, which often lives in and around human homes. The species takes blood meals from a variety of animals, including humans. It is this multispecies sampling that makes *Culex* a vector for West Nile virus, a nasty disease that causes die-offs in birds and encephalitis in humans.[6]

The second species is more troublesome still. *Aedes aegypti* prefer to take multiple small blood meals from humans and other animals, earning them the nickname "snack biters." Like *Culex*, *Aedes* mosquitoes are an important vector of human diseases. They carry dengue, chikungunya, and yellow fever, among others. *Aedes aegypti* are also especially difficult to control. The best way to limit their numbers is to empty the pools of standing water where they lay their eggs—a task far easier said than done. Cisterns, plant drip trays, bird baths, kiddie pools, and even discarded litter are all common breeding sites for *Aedes* eggs to hide. Puddles of water as small as those contained in a bottle cap are a risk. Even ornamental plants can become sites for *Aedes* to lay their eggs.[7] And as if rooting out all sources of standing water were not hard enough, the eggs can survive months of desiccation and even mild winter temperatures.[8]

Back at the town hall, Doyle marches through slide after slide of detail on everything his organization does to keep the local mosquito population at bay. To combat *Aedes*, the Florida Keys Mosquito Control District runs a complex and expensive program to find and eliminate breeding sites. Nine inspectors make rounds to examine thousands of properties for standing water. If they had enough resources, they would visit every house every seven to ten days, Doyle explained. But budget constraints mean they only inspect each house once every month or so. Year-round educational campaigns enlist residents to do their part by maintaining water-free properties. Fumigation teams, in turn, blanket yards in neighborhoods with large concentrations of mosquitoes with pesticide fog. The district pays for helicopters to spray the area with larvicide to the tune of $25,000 (in 2012 dollars) every two weeks during the rainy season. There is even a program to breed and release larvae-eating fish into abandoned backyard pools and hot tubs.[9] These efforts cost taxpayers over $1 million per year, or about 10 percent of the county's annual budget.[10]

These efforts are laborious, expensive, and inadequate. In 2011, the year before the town hall, 10 to 20 percent of containers inspected were found to contain *Aedes* larvae. Some weeks later, that percentage stretched to almost 40 percent. Only on three occasions did the district meet its goal of finding that less than 5 percent of containers harbored larvae.

The challenges with mosquito control are compounded by another complication. *Aedes* has been evolving to be resistant to active ingredients in insecticides. One 2020 study found that 95 percent of *Aedes aegypti* were resistant to pyrethroid and 31 percent were resistant to organophosphates.[11] What the district needed, said Doyle, was a mosquito control tool that was more effective, less expensive, and less reliant on chemicals.

The second presentation of the evening proposed a solution. The Florida Keys Mosquito Control District had convened the meeting to seek community feedback on a proposed partnership with Oxitec, a company that genetically modifies mosquitoes for pest control. If approved, Oxitec would conduct a pilot test in the Keys by releasing millions of their engineered mosquitoes and documenting how the wild mosquito populations shrank in response. It would be the first such test in the United States.

Oxitec modifies mosquitoes using synthetic biology, a new class of technologies that offers scientists the means to edit the DNA code to life itself. It works like this: the company alters the genetic makeup of the mosquitoes by introducing a synthesized gene that is lethal to female mosquitoes, along with a gene that makes them glow under a special light (for purposes of identification in the wild).[12] Eggs are placed in just-add-water boxes in a neighborhood. A few days later, only the males emerge as adult mosquitoes because the self-limiting gene prevents females from surviving to adulthood. Male mosquitoes do not bite, so releasing only the nonbiting sex means that the genetically modified organism is unlikely to come into direct contact with humans. The modified male mosquitoes are then released into the wild, where they live long enough to reproduce with wild female mosquitoes.

It is in the next generation, in the offspring of the modified male and the wild-type female, where the population is suppressed. All of the pair's offspring inherit a lethal gene that prevents the females from surviving to adulthood. The male progeny are unaffected by the self-limiting gene, so they can survive to adulthood and will inherit the gene that is lethal to females. This process repeats, shrinking the population of subsequent generations of mosquitoes and shrinking the overall population. Done at scale, according to some field studies run by the company, the population of wild mosquitoes declines up to 95 percent.[13] Eventually, these effects end. Over time, the wild-type mosquito population rebounds, returning the ecology to its natural state. But in the interim, the population of mosquitoes will shrink, along with the diseases they carry.

. . .

IN THE YEARS SINCE OXITEC developed its technology, the field
of synthetic biology has blossomed. Perhaps the most important development is the CRISPR system, an innovation for which Jennifer
Doudna and Emmanuelle Charpentier won the Nobel Prize for Chemistry in 2020.[14] CRISPR leverages a gene-editing system that certain
bacteria employ for their own immune defense. In human hands, the
system can be used to identify specific segments of DNA, for example
those that code certain genetic diseases, and replace it with a new set
of instructions.

Already, it's clear that the impacts of CRISPR will be profound.
Scientists are using the system to engineer bacteria to produce fra-
grances,[15] pharmaceuticals,[16] and even fuels,[17] which proponents hope
will ease supply constraints and lower consumer prices. The technology also has the potential to harden agricultural crops to resist cold or
heat or pests, innovations which may someday increase crop yields
and blunt hunger.[18] These are just a few of the myriad innovations
that this new science will enable.

Medical science, too, will benefit. CRISPR is being used to create
a new class of therapies to treat serious diseases like certain genetic
disorders and cancers.[19] For example, in December 2023, the FDA
approved the first gene therapy to treat sickle cell disease, an inher-
ited blood disorder that causes the body to produce blood cells shaped
like a sickle or half-moon.[20] Instead of passing smoothly through the
body's circulatory system, the malformed cells become lodged in small
blood vessels, causing intense pain and damage to soft tissue. Until
recently, the only cure for sickle cell disease was a bone marrow transplant, which is a painful and risky procedure. Not only is the proce-

dure grueling for the recipient, but finding a suitable donor is all too rare: fewer than 20 percent of patients find a match.[21]

CRISPR has paved the way for a new approach. Scientists can now extract and edit the patient's own bone marrow and correct the single faulty gene that causes the sickling. The patient's native bone marrow is then destroyed through chemotherapy, and the gene-edited bone marrow is implanted instead. The treatment bypasses the need to find a matched donor, which will likely expand the number of people who can access a cure.

And that's just the beginning. More medical advances are expected in the years to come. In the words of the Nobel committee's announcement, CRISPR is "taking the life sciences into a new epoch and, in many ways, are bringing the greatest benefit to humankind."[22]

IN THE CASE OF genetically modified organism (GMO) mosquitoes, the technology, it turned out, was the easy part. The question of "Can we?" was put to rest when early pilots in other countries showed that Oxitec's technology could effectively reduce mosquito populations. But residents of the Florida Keys, for example, were left to grapple with a much harder question: "Should we?"

At the town hall in the Florida Keys, the presentations from Doyle and Oxitec gave way to an open question-and-answer session with the audience. Attendees peppered the presenters with questions about the safety and effectiveness of the modified mosquitoes and about the decision-making process leading up to the proposed release. Residents worried about whether the mosquitoes were dangerous to human health, and whether the modified mosquitoes would become invasive or cause harm to the local ecology. One complained that the county government meant "to shove this technology down [their] throats," without regard to community sentiment. By the end of the meeting, it was clear that the town hall audience was deeply skeptical.

On that question, at least, the audience member need not have worried. Opponents of the release had a powerful weapon on their side: regulatory uncertainty. The US regulatory system is too often regarded as too slow, too cumbersome, too stifling. And there is some truth to those accusations, of course. New rules about, say, tobacco products can take many years to manifest, thanks in part to powerful corporate lobbies. In 2022, an infant formula shortage that required imports of products from Europe took months to resolve.[23] Approvals for new medical devices, or even new sunscreen ingredients, can take years.

But like many facets of public health, the successes are often imperceptible. We generally hear about regulatory issues only when there is a problem. When everything is working smoothly, there is little to call our attention. This is the case with vaccines—when infection rates are low, it is easy to overlook the role that vaccines played in securing that win. Similarly, we trust that our food is safe, and our medicines are effective thanks to the Food and Drug Administration, the CDC, and the United States Department of Agriculture (USDA). Their work is continuous, though we may only perceive it when an issue emerges.

On the other hand, flexibility is not a strength of most government bureaucracies. If you've ever needed something out of the ordinary from, say, the Department of Motor Vehicles, you know what I mean. Anything outside the normal playbook can quickly become Kafkaesque. In the case of GMO mosquitoes, even the federal departments themselves did not know where to begin. The USDA had previously handled review of GMO bollworms, which are used to control pests that attack cotton, corn, and other plants. But Oxitec's mosquitoes were meant to benefit human health, not plant health, so the USDA was ultimately ruled out as the responsible authority. The Environmental Protection Agency (EPA) regulates pesticides, so they could plausibly claim jurisdiction, as could the Food and Drug Administration (FDA), which regulates medical treatments. Sorting out which government agency was the appropriate regulatory authority for these mosquitoes took years.[24]

The FDA eventually took the first pass at the review—a process that took six years—only for the FDA regulators to determine that genetically modified mosquitoes were outside their jurisdiction after all.[25] The matter then went to the EPA, where the labyrinth of assessments and considerations began again. For the next three years, that agency considered whether to permit the trials in the Keys.[26]

The protracted processes proved advantageous for those opposing the pilot program to release the genetically modified mosquitoes. With a regulatory decision pending, the trial could not proceed. But for proponents of the pilot, including the Florida Keys Mosquito Control District, the delays were costly. Each year, $1 million more was sunk into the mosquito control activities that yielded only modest results at best.

Finally, a decision emerged from the vault of overlapping federal deliberation. In May 2020, the EPA gave approval for the pilot program to go ahead. Yet still more hurdles lay ahead. The next port of call was the Florida Department of Agriculture and Consumer Services. For their part, the department collected input from no fewer than seven other state regulatory agencies, including the Florida Department of Health, the Fish and Wildlife Conservation Commission, and the Bureau of Chemical Residue Laboratories.[27]

This arrangement, with seemingly duplicative reviews at multiple levels, is a peculiarity of the federated nature of governance in the

United States. Federal standards typically set the minimum bar for what is permitted or enforced. States must meet these standards (though they don't always enforce them), and in some cases states may opt for stricter standards and regulations, if they wish. This two-tier system can cut both ways, sometimes working to the advantage of public health priorities and sometimes rolling progress back. During the years when tobacco control was advancing, the battle was primarily fought at the state level. Bans on indoor public smoking were first introduced in Arizona in 1973.[28] By 2009, thirty-seven states and the District of Columbia had some combination of smoke-free workplaces, restaurants, and bars.[29] The ability of states to move forward in the absence of federal action likely pushed ahead the fight against tobacco by years. On the other hand, the opposite can also happen, stalling or rolling back progress on public health priorities. States can, and often do, challenge federal regulations on the grounds that they are an unconstitutional overreach that interfere with state authorities.

BACK IN THE KEYS, federal and state approvals finally came through. The last step was the local level, back where it all began. The Florida Keys Mosquito Control District needed to secure support from the local governance council comprised of elected officials who were accountable to community members at the polling booth. Opponents of the release launched a vigorous campaign—an effort that extended far beyond the residents of the Keys. One sign-on letter accrued over 237,000 signatures,[30] a great majority of which came from nonresidents on the islands.[31] A loose coalition of resident-advocates also launched a website and social media accounts protesting the proposed release, some of which accrued tens of thousands of followers.[32]

Despite their active engagement, opponents were fighting an uphill battle against popular opinion. In the months following the town hall, the Florida Keys Mosquito Control District and academic researchers from the University of Arizona surveyed some four hundred Keys residents about their feelings toward the proposed pilot program.[33] Only half the respondents had heard of the project—a number that surely rose as the effort dragged on. Among those who were aware, a solid 57 percent were supportive. Less than 10 percent were opposed and the remaining quarter were neutral.

Curiously, this popular sentiment did not translate to support from elected leaders. The Key West City Commission considered a resolution called "Saying no to mosquitoes."[34] It passed five to two, weakening the prospects of imminent release in the Keys.[35] It was yet another roadblock in a long and winding saga to determine whether and how the new mosquito-control technology should be used.

The Oxitec pilot program's journey through town hall meetings, regulatory reviews, and ballot initiatives is unusual. It's uncommon for those affected by new biotechnologies to have a say in their introduction, at least to this extent. This intensive community engagement and regulatory review contrasts sharply with the modus operandi of most companies producing new technologies: put it out there and hope for the best. The ideal scenario is probably a combination of the two. A thorough engagement process allows for public input, acknowledging the importance of community voices. But a drawn-out timeline can lead to excessive delays, which carry their own potential for harm. A balanced approach, blending public engagement with streamlined processes, may offer the best outcome.

SEVERAL TIMES IN MY CAREER I have had the sense that Mother Nature is sending us a message: don't get too comfortable, unexpected challenges lie ahead. That message rang clearly in 2015, when a virus related to West Nile, dengue, and yellow fever crossed the Pacific Ocean from one island to the next until it made its way to the shores of the Americas. The newcomer was Zika virus, so named because it was first discovered in the Zika forest of Uganda in 1947.[36] In the intervening sixty years, the virus managed a quiet existence, as far as we know, appearing only intermittently in the human population.[37] Some surveys of blood samples found evidence that Zika virus had circulated in parts of Africa and Asia for decades, but it was never recognized as a cause of severe disease. In fact, the first documented outbreak did not occur until 2007, on the island of Yap in the Pacific Ocean.[38]

In epidemiology, there is no such thing as a boring disease. Even diseases that are well studied and seemingly understood still manage to regularly surprise us with unexpected behavior. For example, it was nearly forty years after the discovery of Ebola virus that scientists learned that the hemorrhagic fever virus could be transmitted through semen months after the original infection.[39] Similarly, although monkeypox (now called mpox) has been on epidemiologists' radar since the days of smallpox,[40] it was not widely recognized as a sexually transmitted infection until 2019.[41] Both possibilities came to light only when the diseases escaped small, rural communities and moved to urban areas where they could infect great numbers of people.

There are multiple mosquito-borne diseases that circulate widely in the Americas, and new ones appear with unsettling regularity. Dengue, also known as breakbone fever, surges in epidemic waves that peak every few years. Dengue has the unusual property of sometimes being more severe the second or third time someone becomes infected, whereas for most viruses, an initial infection teaches the immune system to respond more effectively the next time around. In

2013, dengue was joined by chikungunya, emerging in the Americas as another mosquito-borne disease that causes periodic waves of infection. But diagnostic tests showed that the new epidemic was caused by neither dengue nor chikungunya—it was Zika virus disease instead. The most prominent symptoms of this newcomer virus were fever, joint pain, and fatigue. But as the epidemic unfolded, troubling new health effects began to emerge.

By mid-2015, approximately six months after the virus first surged in Brazil, obstetricians noticed an increase in microcephaly, a rare birth defect. Babies born with microcephaly have an abnormally small skull and brain, which is often accompanied by intellectual disabilities and permanent facial differences. In response to the change in trends, the Pan American Health Organization (PAHO), the arm of the World Health Organization covering the Americas, issued an epidemiology alert.[42]

The Brazilian state of Pernambuco, for example, typically saw an average of 10 cases of microcephaly per year. In the first eleven months of 2015, 141 cases were recorded.[43] Similar trends were reported in other Brazilian regions, alarming epidemiologists who first suspected that environmental contamination may be to blame. But soon, an alternate hypothesis began to surface. The European Centre for Disease Prevention and Control speculated that the recent spike of microcephaly cases was linked to the new epidemic of Zika virus, and as the scientific evidence mounted, the link became clearer.[44] By the next year, there was no longer any doubt. Zika virus, known to science for decades as an uncommon curiosity that caused only minor and sporadic isolated cases in Uganda, was not only driving a major epidemic of disease in the Americas but was also causing severe birth defects in babies born to mothers infected during pregnancy.

Zika virus joins a very small list of viruses known to harm developing fetuses. Rubella, also known as German measles, is perhaps the best-known teratogenic infection. Women infected with rubella during the first trimester of pregnancy are at risk of miscarriage or stillbirth.[45] Babies that survive are at high risk of congenital deafness, cataracts, heart defects, and other disabilities. Rubella has been eliminated from high-income countries thanks to a safe and effective vaccine, usually given together in a formulation that also protects against measles and mumps. In the US, for instance, only eight congenital rubella cases were reported between 2016 and 2019.[46] Yet in countries with inadequate vaccine coverage, hundreds of babies with congenital rubella syndrome are born each year.[47]

Another, more common, virus that is known to cause birth defects is cytomegalovirus (CMV), which flies under the radar, rarely causing symptoms in healthy adults (just like Zika). If a mother is infected during pregnancy, however, CMV can cause premature delivery and low birth weight. Babies with congenital CMV can also suffer from birth defects like microcephaly, hearing and vision loss, intellectual disabilities, and congenital heart defects.[48] Unlike rubella, there is no vaccine for CMV, and it circulates widely with little in the way of surveillance or control. An estimated one in two hundred babies in the United States is born with congenital CMV.[49] In developing countries, rates are even higher.[50]

AS THE ZIKA EPIDEMIC expanded in Brazil, alarm in the United States grew. Although mosquito-borne diseases are not as common in the US as they are in South and Central America, some southern US states and territories do harbor the species of mosquito that carries the

Zika virus—the same *Aedes aegypti* that was the subject of so much debate in the Keys.

In July 2016, more than a year after the outbreak, Florida residents got concerning news. The Wynwood neighborhood, a trendy hot spot north of Miami, reported that a case of locally acquired Zika virus disease had been found. Additional reports from around Miami Beach soon followed. The CDC took the highly unusual step of issuing a warning advising women who are pregnant or seeking to become pregnant not to travel to the affected areas.[51] One news report described how, nearly overnight, those neighborhoods became ghost towns[52] (an eerie preview to the closures that shuttered businesses during the coronavirus pandemic four years later).

These domestic cases came at a pivotal time, just as the country was locked in the contentious lead-up to the 2016 presidential election that pitted Hillary Clinton against Donald Trump. The election was already slated to drive record turnout. In Monroe County and Key Haven, Florida, an extra item appeared on the ballot. Residents were offered a nonbinding referendum on the release of the genetically modified mosquitoes. With Zika virus top of mind and the presidential election filling the polls, residents had an unusual opportunity to register an opinion on whether the trial of modified mosquitoes should be allowed.

The measure passed, garnering 57 percent in favor in Monroe County.[53] With green lights from federal and state authorities and support from county residents through the referendum, the path was clearing for Oxitec trials to take place in the Keys—though the process was still quite slow. Final approval came from the Florida Keys Mosquito Control District in August 2020, when the board voted four to one in favor of the trial.[54] A ten-year gauntlet had come to an end.

The test site was a small, unincorporated community of Key Haven, which is among the southernmost islands of the archipelago.[55] Home prices start in the millions, and nearly every home is bordered by water on at least one side. Key Haven had the useful property of being easily split into thirds, making it ideal for studying the effects of the GMO mosquito trial. One third of the area would be used as a release site, and one third would be used as a control site. *Aedes* mosquitoes stick close to home their whole lives, so the middle third would serve as a buffer between the zones to keep the mosquito populations separate. If the Oxitec pilot program worked, there would be substantially fewer mosquitoes in the test zone.

The next spring, boxes containing millions of male Oxitec mosquitoes were placed at six test sites around Key Haven.[56] Once the project was underway, more soon followed—another example of the irrepressible march of progress. The initial pilot program was followed by an expansion[57] that included the release of millions more mosquitoes.[58] The Florida Keys Mosquito District is also piloting the release by Oxitec competitor MosquitoMate,[59] which uses a different, non-GMO technology to reduce the mosquito population.[60] Neither trial has published publicly available results yet, but if Florida observes the same results as Oxitec trials conducted in other parts of the world, they can expect around a 90 percent reduction in the mosquito burden.[61]

THE GMO MOSQUITO experience is a unique window into how regulators and the public regard new public health technologies. The mix of optimism about what a technology promises and fear of what it might unleash will be familiar to anyone following the latest news on artificial intelligence, self-driving cars, virtual reality, or any of the other astounding innovations soon to hit shelves. There is so much good to be had, so many lives to be saved, years to be improved. The pilot program in the Florida Keys with GMO mosquitoes may not have happened had the new epidemic of Zika virus disease not added urgency to the decision, an observation that recalls the role of the anthrax mailings in fortifying the nation's biodefense. But at the same time, the process was far from ideal. Regulatory approvals were extraordinarily slow. In the decade it took for the genetically modified mosquito pilot program to commence in the Keys, the world saw an explosion of Zika. Meanwhile, technology that could help keep mothers and babies safe endured regulatory purgatory and the court of public opinion. How should we regard these trade-offs?

For one thing, it depends on a community's broader context. In Florida, extensive deliberation is sensible, on balance. The Keys are moneyed, and although residents and visitors are at risk of dengue, chikungunya, and Zika in theory, in reality few cases are recorded. Meanwhile, in other parts of the world, the matter is much less academic. In Puerto Rico, a US territory just over one thousand miles southeast of the Keys, the Zika virus epidemic hit hard, with more than forty thousand cases recorded through May 2017. At least thirty-eight babies were born with microcephaly and other Zika-related birth defects; a tally that is likely an undercount.[62] Technology to suppress the mosquito population there could materially change the risk of Zika and other *Aedes*-borne diseases.

Indeed, when it comes to mosquito-borne disease, most of the world looks more like Puerto Rico than the Keys. Yellow fever, a flavivirus carried primarily by *Aedes* mosquitoes with a case fatality rate in the range of 20 to 50 percent, kills some thirty thousand people each year around the world.[63] In 2015, just before the Zika outbreak was getting underway in the Americas, epidemics of yellow fever took hold in Angola, the Democratic Republic of Congo, Uganda, Brazil, Colombia, and Peru.[64] Worse yet, the epidemic coincided with a shortage of the vaccine, prompting the WHO to take the unprecedented step of recommending that fractional dosing, or smaller than normal doses, be used to stretch supplies.[65] And malaria, a parasitic disease carried by dozens of species of *Anopheles* mosquitoes, is one of the most prolific infectious disease killers in the world. Some 619,000 people died of malaria in 2021, 77 percent of whom were children under five.[66] Against these grave and widespread threats to public health, the balance of risks and benefits of genetically modified mosquitoes begins to look quite different than it would in a lightly affected community. I am left wondering: Where are the technological solutions for those communities, and how can they be brought forward at a speed that matches the risk?

ON THE OTHER HAND, some technologies are cast into the world too quickly, and without any governance at all. Not all technology developers are as committed to public engagement as Oxitec. Indeed, Oxitec is a business that cannot exist without diligent attention to regulatory requirements and accountability to the electorate. This is not always the case with new technologies, particularly in the realm of synthetic biology. The pool of people to whom this new alchemy is available is expanding by the day. As with planetary protection, as the number of stakeholders grows, so too do the opportunities for mishaps or misuse.

This democratization stands in contrast to the other high-consequence fields like nuclear security, where the materials are tightly tracked and controlled, and are difficult to pull together without a connection to a legitimate source. Nuclear security is also a constant discussion between governments, where well-established diplomatic relationships and regulatory mechanisms facilitate control. Rare is a home enthusiast, or a villain that could obtain, store, and purify nuclear materials.[67] As an additional layer of protection, nuclear security has regulatory and oversight bodies at both the national and international levels that are empowered to create and enforce safety and security protocols.[68]

In biology, nothing as muscular exists. The current paradigm is more like what has been cobbled together for artificial intelligence, which is now widely available and nearly impossible to rein in. For biological technologies that could be used for nefarious purposes, the primary mechanism for governance at the international level is the Biological Weapons Convention, a fifty-year-old treaty that prohibits the use of biological weapons. It is the only such treaty to ban an entire class of weapons. The convention enjoys near universal support, which represents a remarkable consensus in the fractured world of international relations.

Representatives of states party to the convention convene each December in snowy Geneva, Switzerland. The United Nations complex where proceedings are held is grand. Diplomatic envoys convene in an impressive circular auditorium several stories tall. The walls are covered by warm wood paneling overlooked by windows where UN staff simultaneously translate proceedings into one of six official languages. Seats for nearly two hundred delegations are arranged in a half-moon shape—which makes the presence of a single small coffee kiosk desperately inadequate for all the jet-lagged attendees.

Despite its broad support and long history, however, the Biological Weapons Convention struggles to keep up with the swiftly evolving nature of biological technologies. To be fair, it was never meant to address the challenges posed by democratized biotechnologies in the twenty-first century. Its purpose, both at the time of its inception and now, is to prevent state actors from using biological weapons as tools of war. But for want of other relevant forums, the BWC has become a stilted vehicle for considering other emerging issues, including misuse of biology by rogue groups or lone actors.

One policy proposal illustrates the challenges that the BWC has in moving nimbly. In 2018, I traveled to a small conference in Tianjin, China, attended by scientists and diplomats from around the world. China is a powerhouse of science and technology, and its influence is flourishing, driven by decades of deliberate strategy and investment by the Chinese government.

We were there to discuss the future of biosafety and biosecurity in synthetic biology. In particular, the conference was meant to build support for a code of conduct for biological researchers, an idea that had been circulating in the BWC community since at least 2005.[69] Three years later, China and Pakistan renewed discussions when they introduced a model code of conduct to serve as a draft for discussion.[70] Ten years had passed since then, and the idea still percolated, neither moving toward adoption nor fading away entirely.

Adopting the code of conduct should have been an easy win. It would be a short document setting out what sort of ethical conduct scientific professionals should adhere to. Universities and other scientific institutions would be encouraged to make it part of their curriculum as a sort of honor code. Participation in observing the code would be voluntary. Yet years had passed since the idea first gained traction, and little progress had been made.

Back in Tianjin, after a morning of conference sessions, we all piled on a bus and embarked on a tour of regional biology laboratories. Each stop followed a similar routine. The bus would pull into a nondescript office park or industrial zone. We would file into an air-conditioned lobby, past a security desk, and through long hallways with fluorescent lighting. The corridors were punctuated with picture windows that overlooked huge laboratories filled with scientists in white coats and tennis shoes working at biosecurity hoods. These, we were told, were synthetic biology laboratories where some of the most innovative science in the world was being produced. Back in the lobby, framed covers of prestigious scientific journals like *Cell* and *Science* adorned the walls, homage to the peer-reviewed and award-winning research generated from these laboratories. We visited four such laboratories in a few hours. What we saw represented a tiny fraction of the biological sciences industry in Tianjin, to say nothing of the rest of China. And although China is an especially prominent actor on the world scientific stage, nearly every country in the world has similar work underway.

The juxtaposition of the staid conference room discussions about whether and how to finally adopt a voluntary code of conduct was cast into sharp relief. The future of synthetic biology, I realized, may very well not follow the path set out in their mosquito trials. The future is already here, unfolding at breakneck pace, in China and in laboratories around the world. In public health and biosecurity, we would do well to keep up.

10

MYSTERIES

A year and a half into the COVID-19 pandemic, I was keen to visit my grandparents, who live several states away. Like most families, the pandemic had kept us apart for far too long. As soon as we could all count ourselves fully vaccinated, I boarded a plane (still accompanied by a KN95 mask and with a negative COVID-19 test under my belt) and went for a visit.

The very next day, the sniffles set in. I feared that I'd brought the virus with me, but a second COVID-19 test also came up negative. The flowering trees lining every street made me suspect that seasonal allergies were causing my symptoms, but I couldn't rule out illness altogether. It had been over a year since I'd had a cold. At that point, my kids had been out of school and day care for a year, I had been working from home, masking was near universal, and I had become accustomed to a germ-free life. I had nearly forgotten that such a thing as a cold was possible. After a long period of good health, in a cruel twist of fate, I had succumbed to sniffles while visiting my elderly grandparents.

Colds are sneaky. What young and healthy people experience as a minor nuisance can pack a big punch for people who are very old or very young. Little kids, having not yet encountered the many and varied viruses that circulate among us, are prone to getting sick from every pathogen they encounter. This can be especially dangerous for infants because their immune systems are not yet strong enough to mount a proper fight. They can become very ill from a virus that would barely touch an older child. Later on, as children grow older and stronger and their immune systems become wise to everyday bugs (and as they become less prone to putting everything in their mouths), they become sick less often. When older children and healthy young adults do fall ill, they usually recover just fine at home. But later in life, vulnerability to infection returns. Bodies and immune systems grow frailer, and conditions like diabetes or heart disease make it harder to fight germs off. Common pathogens again pose considerable threat. Before the COVID-19 pandemic, in adults aged eighty-five and older, for example, the death rate from pneumonia and influenza was over seven times higher than for people ages sixty-five to seventy-four.[1] COVID-19, respiratory syncytial virus (RSV), influenza—all can be especially severe in older adults.[2]

That was precisely my fear when visiting my grandparents. Although I had only the sniffles, and several negative COVID-19 tests to boot, I had no way of distinguishing what was causing my congestion. And without that information, I had no way to make an informed decision about how to keep them—and the other people around me—healthy.

Thankfully, my grandparents did not become ill from my visit, but the experience—particularly against the backdrop of a historic respiratory pandemic—was a reminder of just how many epidemiological mysteries we live with. Some days it feels to me that, for all that scientists have learned since the days of Dr. John Snow and the dawn of epidemiology, there remain more questions than answers about the comings and goings of viruses. The truth is, we know little even about what plagues us day to day.

Take the coughs and sniffles that most of us encounter each winter. What we refer to collectively as "colds" are a motley collection of rhinoviruses, enteroviruses, adenoviruses, coronaviruses, and dozens of other germs.[3] A sufferer is very unlikely to know which he carries. The symptoms from one to the next are virtually indistinguishable. And although diagnostic tests exist that can tell them apart, doctors rarely use them because there is not much that can be done regardless.

The forces behind the seasonal nature of these respiratory viruses are also somewhat mysterious. Each winter there is a resurgence of influenza, RSV, and other pathogens that cause cold and flu-like symptoms. This is a pattern seen annually in both hemispheres.[4] And yet, little is understood about what role winter plays, exactly. Researchers have found that lower relative humidity improves viruses' ability to survive on surfaces for longer periods of time, thus making it more likely that a human will pick it up.[5] But this observation alone is unlikely to explain the strong seasonal pattern, as humidity levels during winter are more variable than patterns seen in respiratory viruses.

The virus that causes COVID-19 likes winter, too, but it has also been active in late summer, and the reason for this is unknown.[6] Epidemiology lore has it that harsh winter weather and hot summers both drive people indoors, where they spend more time in close quarters in poorly ventilated spaces. But this too does not feel like a sturdy explanation. Most people in the United States live in climate-controlled buildings[7] year-round and spend a minority of their time outdoors.[8] I'm skeptical that indoor life is solely to blame.

I am fascinated by these everyday curiosities. There is so much to learn about epidemics and how to control them, even after all these years. It is those puzzles that keep me passionate about my profession. But it's another set of mysteries—ones that hint at the return of past dangers—that leave me restless at night.

ONE OF THE MOST ENDURING enigmas in epidemiological history is a disease that disappeared as mysteriously as it arrived. Its symptoms were bizarre, and its effects on victims were permanent. Yet to this day, over one hundred years later, scientists do not know what caused the epidemic, how it emerged, and why it left with hardly a trace.

The epidemic, known then as "sleeping sickness" to the layperson and "encephalitis lethargica" to scientists, was first recognized in Vienna, Austria, in 1916, and it unfolded alongside the 1918 influenza pandemic—both likely accelerated by the movement of troops in World War I.[9] Cases were seen around the world, including across Europe, North and Central America, and India.

The epidemic—though even this label is guesswork—hit its victims with a one-two punch. Dr. Constantin von Economo, a wealthy Greek-born Austrian neurologist of "independent means" (owing to his noble title of baron and, later, his marriage into the wealthy Schönburg family) was the first to characterize the debilitating disease.[10] In the acute stage, victims experienced the same flu-like symptoms that herald the onset of many infections, including fever, fatigue, headache, and vomiting. The illness then progressed into a curious array of neurological symptoms, leading one modern scholar to speculate that women persecuted in the Salem witch trials were actually victims of encephalitis lethargica.[11]

Von Economo identified three main acute forms of the disease. In the somnolent-ophthalmoplegic form, patients were excessively sleepy—although some people who were thought to be asleep later reported that they were aware of what was happening around them. Sufferers had trouble controlling their eyes, which sometimes became stuck while rolled back in their heads, among other neurological symptoms.[12] This form of encephalitis lethargica was the most common form and also the deadliest, killing over half of victims from respiratory failure caused by brain swelling.[13] In the second, hyperkinetic form, patients had difficulty controlling their movement. They made noises involuntarily and suffered spastic twitches, jerks, or writhing. Victims of the hyperkinetic form alternated periods of muscular problems with periods of fatigue and sleepiness, sometimes accompanied by delusions.[14] And the third form, amyosatic-akinetic, was least common, and perhaps most striking in its presentation. Victims appeared excessively stiff, hardly able to move their locked muscles. If physically manipulated by someone else, though, they could be repositioned or compelled to move without much resistance. Physicians described their form as waxy—rigid yet pliable when manipulated.[15]

Following the acute phase, those who survived their illness would usually recover their functioning. But for some victims, the return to normal was only temporary. Years or even decades later, late complications would develop into a chronic form. Difficulties with movement, speech, and cognition returned.

Encephalitis lethargica is not the only disease that can flare decades after the initial infection. Polio victims can develop progressive muscle weakness later in life,[16] and the virus that causes chicken pox can later reappear as shingles.[17] There is even emerging evidence that Epstein-Barr virus plays a role in triggering multiple sclerosis.[18] But the chronic form of encephalitis lethargica was unique for how thoroughly it could incapacitate its victims.[19] Although conscious, victims were catatonic, seemingly without cognition. As famously documented by author and neurologist Oliver Sacks in *Awakenings*, people stricken by the chronic form of encephalitis lethargica existed as if in suspended animation. Only with considerable prompting could they be convinced to move, eat, or participate in daily life.

THE EPIDEMIC receded as mysteriously as it began. After a peak in 1924, the number of new cases gradually dwindled to nearly zero. (Though sporadic, one-off cases that seem similar to encephalitis lethargica are diagnosed and reported from time to time.[20]) And now, a century later, memories of encephalitis lethargica are receding further into history. There are few (if any—there are no comprehensive records) survivors still alive and the cadre of clinicians who cared for them are aging. Will we ever learn what was behind the epidemic that swept through after the First World War?

Experts have speculated about various viruses and bacteria, as well as noninfectious sources like toxins. So far, only one of those ideas has had staying power. Because the epidemic flared closely on the heels of the 1918 influenza pandemic, some scientists have theorized that the disease was related to flu, just as long COVID follows SARS-CoV-2 infection. However, little evidence has emerged to support that link. Stored samples of tissue taken from people with encephalitis lethargica found no trace of the influenza virus, and epidemiological studies looking for associations have not turned up any compelling results.[21]

So complete is the mystery that some experts wonder whether the epidemic was a single disease at all, or whether several different disorders were wrongly lumped into a single diagnosis.[22] Clinical observation—a doctor's keen eye—was the primary tool for making a diagnosis in the early twentieth century. It's not a stretch to wonder if rumors of an epidemic of neurological disease led clinicians to settle for a diagnosis of encephalitis lethargica for their more mysterious cases.

Whatever the cause, the epidemic vanished for no discernible reason. It is one of the few in history to do so spontaneously. We are rarely so lucky. Victories against epidemics are usually hard fought, and even the best efforts sometimes do not yield the desired results. That's why I find the retreat of encephalitis lethargica to be a haunting reminder of just how little we know about the secret lives of outbreaks.

THE SPONTANEOUS resolution of the sleeping sickness is noteworthy in part because it is rare. Less rare, unfortunately, is a pathogen's spontaneous emergence. I would not be surprised to learn that there are dozens, if not hundreds, of outbreaks each year that go undetected, including those caused by previously unrecognized viruses. Most of these outbreaks are small and they sputter out on their own after just a few infections—but each could potentially ignite an epidemic.

Even in hospitalized patients, it is not uncommon for people with severe respiratory illness to test negative for the most common viral pathogens.[23] Physicians frequently opt against seeking a precise diagnosis, instead offering supportive care based on symptoms, because sequencing tests are expensive and unlikely to change the treatment approach.

The mystery of what viral varieties circulate in humans is even more pronounced outside of hospitals. Most people with respiratory symptoms recover at home without interacting with the medical system, like I did with the stuffy nose I developed while visiting my grandparents. We all know that there isn't much point in visiting the doctor if we're slightly under the weather and otherwise healthy. A virus that causes mild symptoms could easily circulate undetected for quite some time. This makes it difficult for epidemiologists to identify novel strains. It is primarily the responsibility of astute clinicians, like Dr. Carlo Urbani with SARS, to notice unusual patterns in the patients they are seeing and to bring that to the attention of public health authorities.

Some experts would like to redouble efforts to identify possible outbreaks before they even begin. If it were possible, the thinking goes, to identify dangerous new viruses in animal populations before they become established in humans, it might be possible to stop them before they cascade into the next pandemic. To that end, researchers have launched multiple "viral discovery" projects to document what circulates out of sight.[24] Virus hunters venture into the wilderness to take saliva or blood samples from animals, ranging from bats and primates to pangolins. This approach appeals because it combines Indiana Jones–style scientific adventure with the prospect that we could someday see an outbreak coming before it ignites.

These endeavors have yielded a catalog of thousands of viruses, including many previously unknown to science. Some of the viral sequences are eyebrow raising, bearing remarkable similarities to the viruses that cause SARS, MERS, and COVID-19.[25] Other surprises include the discovery of familiar pathogens found in places where they were not known to circulate, suggesting that those locations are vulnerable to outbreaks.[26]

But the promise of being able to look for threats so far upstream presents a curious problem: With so little information, how are we to know what actions to take? There are many animals, many viruses, and few clues about how to make sense of it all. Sequences on their own reveal little about a pathogen's potential. They do not necessarily reveal whether a pathogen can infect humans, how easily it spreads, or whether a pathogen is particularly deadly. Comparing novel viral sequences with known viruses allows scientists to make some educated guesses, but the devil is in the details. A small number of mutations in a sequence can materially change how the virus behaves.

The troubles do not end at our inability to assess the risk a virus might pose to humans. We also cannot say anything about when or where an animal-based virus might make the leap to humans beyond acknowledging that the two must come into contact at some point. Scientists know precious little about the epidemiology, microbiology, anthropology, and ecology of how, why, and when new viruses emerge. Without further context that can place the viral sequences into a broader pattern, simply knowing that a virus exists in the wild is of little use.

And even if we were to know the answers to these questions, what to do about it remains unclear. Some proponents of "premonition" inspired programs say that if we know a virus exists in the wild, public health authorities would be better prepared to respond quickly should it emerge in humans. I'm not convinced. While having an early notice to be on the lookout might make diagnosis clearer or faster, that alone is not a tool for stopping transmission. Whether a virus is new or previously recognized, whether it was foreseen or diagnosed late, the same tools that stopped SARS will be used again to break chains of transmission: contact tracing, isolation, and quarantine. If the epidemiological response is the same, then what, specifically, is the value of beginning the search in animals?

Some of my colleagues would argue that getting a sneak peek allows scientists to get a jump on preparing another powerful public health tool: vaccines. Unlike contact tracing, isolation, and quarantine which are initiated only after an outbreak is underway, vaccines can prevent people from becoming sick in the first place. It stands to reason that we would want to call upon them as quickly as possible. In recent years, the goal has been to develop the infrastructure to create a novel vaccine in one hundred days.[27] The trouble is that developing a vaccine is an enormously difficult and costly undertaking, with research and development costs starting in the hundreds of millions of dollars and going up from there.[28] It is impractical to expect that a vaccine will be developed for viruses that pose only a theoretical risk. There are plenty of diseases which infect humans, including those that infect many people and cause severe illness, for which there is no vaccine. For many infectious diseases, there is not enough of a market to justify the research and development costs. Moreover, because outbreaks flare and then retreat in ways that are difficult to predict, researchers often struggle to enroll enough patients in the clinical trials needed for licensing.[29] Those obstacles would hardly be improved by adding a pathogen for which there are no known human cases at all.

Last but not least, there is the matter of when to begin the clock on a one-hundred-day vaccine. Any epidemiologist will tell you that sooner is better when it comes to applying interventions, but so too will they tell you that crying wolf is poisonous to a field that is chronically underfunded and often in precarious standing with policymakers. The development of any vaccine is eye-wateringly expensive. An accelerated process is likely to be even more so. How will public health officials know when to activate the emergency vaccine capability?

To consider what it means to assess whether an outbreak is a sufficient threat to justify great cost, consider this thought experiment.

Imagine that a new virus emerges that has never been seen before, and that you are the epidemiologist charged with leading the investigation.

A local physician calls your office to report that a man has died of an infection, and that his wife shows similar symptoms. The hospital laboratory was not able to identify the pathogen, and the treating physician has a bad feeling about it. As the epidemiologist in charge, you have two immediate priorities. First, you must characterize the outbreak by assessing how deadly the pathogen is and whether (and how quickly) it transmits from person to person. Second, you must prevent the virus from spreading further. You set about these tasks immediately.

You begin by learning more about the two suspected cases, the man and his wife. You quickly discover that they live in the same household. This gives you little to go on to assess transmissibility. Household members could infect one another through any route of transmission. One spouse could have infected the other—a pattern that suggests human to human transmission—through skin to skin contact, sexual transmission, or even aerosol transmission. Each of these possibilities carries very different risks in terms of pandemic potential. Sexual transmission is least likely to result in a pandemic, at least not a fast moving one. Aerosol transmission, on the other hand, bodes very poorly because it suggests that the virus can move easily between people, even if they are distant from one another. But that's not all. Because the husband and wife live in the same home, they could have both become infected from an environmental source, like an infected animal. If that is the case, the pathogen may not be capable of spreading further. With the data you have in hand, there is no way to determine which route of transmission is most likely.

Through a series of interviews, you make copious notes about all of the people, places, and environments the couple has come into contact with in the last month. There is nothing to learn from this information now with so few cases, but if the outbreak continues to grow, you hope to use that data to detect a pattern.

Next, you set about contact tracing. Let's assume that your concern is great enough that you ask the people who have been in contact with the couple to remain at home in quarantine and to monitor themselves for symptoms. Your staff will call them each day to check in. This serves two purposes. First, it addresses your second goal of preventing the virus from spreading further. If any contacts become ill, the quarantine will ensure that they do not pass it on. Second, the careful monitoring of contacts will teach you how the virus behaves. If none of the contacts develop symptoms, that suggests that the pathogen, whatever it is, is not particularly transmissible. On the other hand, if contacts do develop symptoms, you will learn from their journey through illness, including the incubation and infectious periods, two critical elements of disease transmission that inform how epidemiologists must respond. But for how long must the contacts remain in quarantine? The incubation period of infectious diseases usually ranges from a few days to one or two weeks (though in some cases months or even years, in the case of rabies or prion diseases). Your risk assessment of this pathogen will proceed at that pace.

Meanwhile, you turn your attention to severity. The husband has died, and his wife is quite ill, suggesting that severity may be high. Although you have only these two data points, this puts your new virus in the same category as hemorrhagic fevers like Ebola, for example, that have an average case fatality rate of 50 percent.[30] On the other hand, you learn that the husband had several serious health conditions. In fact, his immune system was suppressed because he was undergoing cancer treatment. Perhaps it was not that the virus is particularly deadly, but that he was medically vulnerable. You will have to collect more data to know—but that will take even longer. Not only will you have to wait for more people to become ill, but you will also have to wait for them to progress through the course of their illness. Once an infection has produced symptoms, it can take days or weeks to progress to either recovery or death, and for that duration you will be observing and collecting data. This cycle of watching, waiting, and learning will unfold several times before you form a sketch of what you are facing.

Unfortunately, time is not on your side. Exponential growth is a deceptive but ferocious force. To the uninitiated, early exponential growth can impart a false sense of security. At the beginning of an outbreak, case counts are small and seem to grow slowly. In an outbreak that doubles every five days, for example, two cases become four and then eight over the course of two weeks. Not so bad, right? This is an illusion that can continue for weeks or even months. Then, in what feels like a sudden transition but actually follows the expected trajectory of exponential growth, numbers explode. In the meantime, what would you advise the policymakers who are asking if they should hit "go" on the vaccine machine? The costs of inaction against a would-be pandemic are great, but so too are the costs (both monetary and reputational) of crying wolf.

Meanwhile, another problem emerges: one hundred days is a long time. It is equivalent to fourteen weeks or over three months. We live in an age where international air travel is more accessible and affordable now than at any other time in human history. According to the International Civil Aviation Organization, in 2019 some 4.6 billion passengers flew on domestic or international flights, which would amount to approximately 1.2 billion passenger trips in a one-hundred-day period.[31] It is not hard to imagine how a virus could make good use of those one hundred days by seeding communities around the world.

Consider the start of the COVID-19 pandemic, the very event that has inspired the creation of this ambitious goal. Within one hundred days of the virus's emergence in China, the cities of Wuhan, New York City, and the Italian province of Lombardy were already in the throes of a crisis.[32] Hospitals were overrun, routine medical care was paused, schools and businesses were shut down, and improvised morgues were needed to manage the deaths. The virus had already become entrenched in dozens, if not hundreds, of other places around the world, firmly establishing it as a pandemic. Developing a vaccine in fourteen weeks would not have prevented the health-care systems in those places from becoming overwhelmed and it would not have spared other countries from needing to respond. Those events had all been set in motion by day one hundred, and even a "just-in-time" vaccine could not have rolled that back. To be clear, had a vaccine been available in the months following the emergence of SARS-CoV-2, it would have saved many lives, but it would not have been enough to avert the crises in China, Italy, and the United States, and it will not be enough to reverse the tides of whatever pandemic comes next.

Defense experts have a saying: You go to war with the army you have, not the army you might want or wish to have at a later time.[33] In the first months of a new infectious disease threat, it is the "staple" public health measures of contact tracing, isolation, quarantine, health education, and infection control in hospitals that will comprise the frontline forces. When a vaccine becomes available it will be a welcome reinforcement, but it cannot be our only plan. So before getting too experimental, perhaps we should take a closer look at what lurks around us day in and day out. Searching for new methods to combat the next pandemic is a worthy goal, but so too is understanding mysteries closer at hand. Or better yet, let's do both, so that we can be innovative without losing sight of proven methods.

In the end, what we need to improve our ability to prepare for, detect, and respond to outbreaks is not any one thing. There are dozens of opportunities on the table, and we must seize them all if we are to reduce casualties from the next pandemic. And in order to bring about those positive changes, we must not overlook the importance of training and funding the next generation of public health experts, the ones who will conduct the investigations that enable leaders to hit "go" on a just-in-time vaccine and then turn that vaccine into a vaccination by getting shots in arms. This work of traditional epidemiology cannot be allowed to fade behind the brighter stars of biomedical technologies. We need all of the above and so we must invest accordingly. The fate of millions of people depends on it.

PROGRESS

For better and for worse, the COVID-19 pandemic revealed a lot about the inner workings of public health. Epidemiologists became household names. I overheard terms like "superspreading events" and "social distancing" in conversations at the grocery store and on the nightly news. Recommendations for brands of face masks and luxury hand sanitizer appeared in magazines. For the first time in my life, the twists and turns of fighting a respiratory pathogen became part of everyone's daily routine.

This experience highlighted how accustomed we are to the many benefits that public health affords us. It is a truism in the field that those wins are invisible when things are going right. We enjoy clean water that is safe to drink, vaccines that stave off infection, and access to the prenatal care that supports healthy pregnancies. Those successes do not make headlines or inspire congressional inquiries. Instead, we see them in the faces of children who reach adulthood, and in the fortunes of people who do not know the scourge of smallpox. It is only when a crisis emerges, like when an outbreak becomes an epidemic, that public health breaks through into the public's consciousness.

And break through, it did. In the first months after the pandemic reached US shores, the state of affairs was dire. As the virus took off in New York City, hospitals became so overwhelmed that special trailers were brought in to serve as temporary morgues.[1] When even that did not suffice, a massive marine terminal warehouse, once used to store debris from the September 11th attacks, was converted into a freezer facility that stored hundreds of bodies.[2] One of the US Navy's hospital ships, the 1,000-bed USNS *Comfort*,[3] was deployed to the New York Harbor, and the Javits Convention Center was converted into a makeshift field hospital.[4] Variations on those harrowing scenes played out in Wuhan, China, and Lombardy, Italy.

To blunt the unfolding crisis, officials turned to community-wide stay-at-home orders to "flatten the curve."[5] The extraordinary measures of closing school buildings, shuttering businesses, and pausing public life were meant to slow transmission enough to prevent New York City's fate from befalling the rest of the country. Compelling individuals to act in a certain way is far from ideal, and the power is used only rarely. Mandatory measures like enforced social distancing are a last resort, to be used only when absolutely necessary to safeguard the broader community's health. At that point, there was little else to be done. Easy access to diagnostic testing was still months away, and vaccines and effective treatments further away still.

Despite the hardships they caused, most Americans recognized the need for these steps. Early in the pandemic, surveys show that the majority of people, at least 85 percent, reported wearing a mask in public all or most of the time in the previous month[6] and some 68 percent of poll respondents reported being concerned that restrictions would be lifted too *early*.[7] This collective action is the single greatest example of solidarity and unity of purpose in my memory, and it had the intended effect of saving lives and slowing transmission enough to prevent the health-care system from collapsing while also buying time for countermeasures like diagnostics to be developed.[8]

But over time, the public health, political, and cultural environments all changed. Children were kept out of school for weeks, then months, and for some—including mine—over a year. Prolonged unemployment pushed vulnerable households into poverty. Families missed out on births and deaths of loved ones.

Those pandemic control measures, though widely adopted in the spring months of 2020, planted seeds of mistrust and anger in people who felt their lives and livelihoods were more disrupted by the restrictions than the virus itself. For some, those feelings matured into a broader backlash against public health. State governments began efforts to roll back legal powers granted to public health departments as long as a century ago,[9] hecklers harassed public health workers,[10] and more people began to question the safety of vaccines (of all kinds, not just COVID-19).[11]

These two tendrils—stewardship of public health during the early days of the pandemic and eventual discord—share the same root, one that constitutes another force in public health that we must confront because it has the capacity to produce both great benefit and harm: us.

As members of the public, we are both the beneficiaries of public health and also the vehicles for either change or stasis. From quelling outbreaks to preventing heart disease, not much progress on health issues can be made without at least some participation from each of us.

For the most part, what is best for individuals and the collective are aligned. After all, you would be hard-pressed to find someone opposed to, say, reducing diabetes deaths. However, the strategies proposed for how we might achieve those goals—like raising taxes on sugary beverages, making school lunches healthier, or requiring masks—often spark controversy among the public and even within the public health community. This contention, made worse by the prolonged upheaval of the pandemic, has evolved into a schism in American life that threatens to set back public health progress and erode our ability to fight future pandemics.

This ongoing friction underscores a fundamental aspect of public health: its reliance on persuasion. Unlike in emergencies, where mandates might be necessary, public health initiatives typically rely on engagement and education. During the smallpox eradication campaign, most people were not required to receive the vaccine. They chose to, because public health practitioners and local leaders persuaded them that it was the right thing to do. Through the SARS pandemic, most people who were infected cooperated with contact tracers because they knew they must in order to protect others. The same is true for those who entered quarantine because they had been exposed. The vast majority did so voluntarily. Residents of the Florida Keys are actively participating in decisions about whether to use genetically modified mosquitoes for vector control. Without their support, the project could not proceed. And just as many people take shelter from storms based on heeding warnings from meteorologists, so too will members of the public if given similar advice by epidemiologists.

Effective communication is therefore the main challenge facing public health practitioners, and so the question becomes: How can we best convince people to adopt healthy practices? The solution requires a nuanced approach, one that relies on listening as much as telling. The relationship between epidemiologists and the public must be a two-way street, where understanding community needs and concerns is as vital as providing guidance and information. When public health efforts fail to nurture community engagement, those efforts inevitably stumble.

We have a lot to learn about how to do that well. Persuasion of any sort is no easy task, and I find that to be doubly true when it comes to influencing habits. Between inertia of long-held habits, personal freedom, and structural forces, there is a lot to overcome. Cautionary tales of failed efforts to change behavior abound. Of all the foibles that beset humans, our inability to consistently match our behavior to our good intentions is perhaps the most maddening. What the wellness industry brightly calls lifestyle change is not so straightforward. My thing is exercise programs. I am perennially launching and abandoning new routines. It turns out I like the planning more than the doing. A 5:00 a.m. alarm followed by a rush of endorphins sounds like a better idea in abstract than in the cold, dark morning.

This applies equally to the struggle to save money, eat better, quit drinking, learn a language, or any of the other aspirations that promise to shape us into a better version of ourselves. As many New Year's resolutioners can attest, shedding well-worn habits and adopting new, virtuous ones is not an easy task. My resolution for the last five years has been to keep a record of every book that I read. Not to read more often, not to read different material, not to examine character arcs or motifs. I only wish to write down the title when I finish. The furthest I've made it is March.

It is not for lack of trying. There exists an ecosystem of programs, apps, and coaches to help people to change their behaviors and achieve their goals. None work as well as we hope. The Cochrane Library, a database of analyses that reviews and synthesizes entire bodies of scientific evidence, is littered with disappointing results. A support group or individual counseling meant to help smokers to quit was proven to work only modestly.[12] Efforts to convince people who work near pathogens, dust, fumes, and other hazards to use protective masks consistently all flopped.[13] Programs to encourage people to exercise more,[14] sit less,[15] drink less alcohol,[16] choose healthier foods,[17] floss—none produced more than a modest change in how people behaved.

It gets worse. Not all public health advice is as clear-cut as the case against smoking or for workplace safety or for exercise. Varied guidance abounds on what pregnant individuals, for example, should or should not eat, drink, or be exposed to. Specific recommendations can be contradictory and often change as the scientific evidence evolves, much to the consternation of women who find that keeping up with the latest recommendations can feel like a part-time job. Parallel industries have sprung up to help new moms either adhere to or unburden themselves from the anxiety of optimizing their health and the health of their baby. Formula was once seen as superior, which gave way to the "Breast is Best" era[18] when epidemiological studies found that breast milk helps to boost babies' immune systems,[19] among other benefits. That approach became a source of unhappy pressure and stigmas[20] for women who can't or don't breastfeed, for whatever reason. Now public health advice has pulled back and landed somewhere in the middle.[21]

Moreover, even when the best way forward is clear, sometimes public health is clumsy at making the case. Health history is littered with well-intentioned ideas and campaigns that failed to launch. Sometimes this is because the messaging has gone wrong. Take, for instance, a 1990s public health campaign aimed at discouraging teenage tobacco use,[22] which attempted to point out the manipulative marketing tactics of tobacco companies through dramatized, fake advertisements. The message was a flop. Over a third of viewers who saw the TV segment thought it was meant to encourage, rather than discourage, smoking. Or an ill-advised campaign by Cancer Research UK, which featured several advertisements that likened obesity to smoking.[23] One advertisement depicted a cigarette pack filled with fries, while another used a design resembling cigarette packaging, replacing the brand name with anti-obesity messages. The intent was to underscore the often-overlooked link between obesity and cancer, but they were roundly criticized as stigmatizing.

There are successes to learn from as well. Cigarette smokers have a life expectancy that is over a decade shorter than people who have never smoked.[24] Consequently, public health officials have been working for years to help people stop using tobacco and to keep new users from taking up the habit. This mostly worked: in the 1960s, over 40 percent of American adults smoked tobacco, by 2023, it was down to 11 percent. A resounding success—but still, one in ten Americans smokes.[25] These setbacks and partial wins have taught me that victories in behavior change are won in inches, not miles.

There are deeper structural factors that influence our health behaviors, too. During the pandemic there were staggering differences in the vulnerability of different racial and economic groups. In the United States, as in many countries, it was the essential workers, many of whom were people of color, who did not have a choice to stay home.[26] It is they who bore the brunt of the pandemic. Disparate health outcomes have always existed, but it was not until the pandemic shined a spotlight on the issue that health inequity began to get the attention it deserved. Unfortunately, it is deviously difficult to solve, with roots ranging from the expensive and inaccessible health-care system to inadequate social supports to failures in cultural competency among health-care workers. I fear that among the issues the pandemic brought attention to, health inequity will be the quickest to fade from our collective consciousness.

These challenges emphasize the extent to which progress in public health is a team effort. Millions of individual choices, big and small, shape the health of our communities. That means that as members of the "public" in *public health*, each of us plays a vital role. Our consistent, small contributions of staying home when ill, getting vaccinated, driving safely, and the myriad other choices we each make are the building blocks of public health's overall success. Those tiny choices are fundamental to the ongoing pursuit of a healthier society for everyone.

WITH THE PANDEMIC now settled into familiarity, we must now turn our attention to the future, understanding the lessons learned and the need for preparedness in the face of new health threats. Mother Nature is a formidable force, and not all natural disasters are preventable. Still, the prevailing sentiment coming out of the pandemic is that nothing like it should ever be allowed to happen again. Certainly, as an infectious disease epidemiologist, that is my aspiration. As experts have long foretold, the impacts of a pandemic are as deep and wide as other major threats to national security like war and famine. We must treat pandemic preparedness with the same seriousness of purpose as we regard those traditionally recognized threats.

The tough reality is that we will almost certainly be confronted with new opportunities to demonstrate our resolve. Although COVID-19 was the first event of its size since the 1918 pandemic, in my experience outbreaks serious enough to require a large-scale mobilization of public health attention and resources occur about every two years on average. And if anything, the pace at which we face these types of threats is accelerating. Close on the heels of the COVID-19 pandemic was the mpox epidemic, followed by a large Ebola outbreak in Uganda.[27] In parallel, several epidemics of cholera flared, including a re-emergence on the island of Haiti, multiple countries in the horn of Africa, and parts of south Asia.[28] Between each of those are smaller outbreaks that must also be tended to, including some that have the potential to grow into a much bigger threat.

This deluge of new outbreaks, which some commentators have dubbed the "Pandemicine" or age of pandemics, will undoubtedly challenge us time and again in the years to come.[29] It is the actions of epidemiologists and elected leaders—and the behaviors of the people at risk—that decide the trajectory of those outbreaks. It is up to all of us to strengthen the defenses against those future challenges.

A final warning: preparing for the next public health crisis requires humility above all else. The next time we face a pandemic we may well be dealt a different hand. In all likelihood, we will be surprised by the exact nature of the threat. And with the advent of new technologies, the form of that threat could be different from any previously seen.

To be ready for the next pandemic, we will need new ways to stay ahead, and the stories, achievements, and cautionary tales of the foregoing chapters highlight the challenges and opportunities of epidemiology. They give us a common way to make sense of often-overlooked work that, when successful, makes it seem as though nothing happened. I contend that these tales and their logic help us to make notes on progress. Beyond any single innovation, it's my view that continual progress is the ultimate win—bigger than any glorious victory—in the arena of public health.

Despite the bumps and twists and turns of the art of progress in the stories I have shared in this book, I hope it is clear that my optimism for public health is untempered. On any great journey there are obstacles. Nothing momentous can be achieved without first overcoming great challenges. The history of progress in public health is no different, and we should expect to face even more problems ahead. But it is also clear to me that in the end, we eventually prevail. We may not always get the exact outcome that we set out to capture, and often not on the timeline we might have hoped. Yet even when we try and fail, our failures still manage to move the needle on staving off preventable death and disease.

ACKNOWLEDGMENTS

So many people made this book possible, and I owe each of them a thanks. Undertaking this project with three small kids at home is no small feat, so I owe a debt of gratitude to my husband and our three girls. Their support has been infinite.

I also could not have done this without my "book team," including my research assistant Anna Bezruki, my agent Margo Beth Fleming, and my two editors, Wendy Wolf and Terezia Cicel. I also drew on the expertise and wisdom of many colleagues, including Stephen Ostroff, Gigi Gronvall, Tom Inglesby, Alison Kelly, Joel and Charlotte Gaydos, Charles Hoke, Farzad Mostashari, Mark Dybul, Lauren Sauer, the Oxitec team, and many others. All have been generous with their time.

And a final note from the DoD: The views expressed in this publication are those of the author and do not necessarily reflect the official policy or position of the Department of Defense or the US government. The public release clearance of this publication by the Department of Defense does not imply Department of Defense endorsement or factual accuracy of the material.

NOTES

INTRODUCTION: WELL-BEING

1. Basil C. Tarlatzis et al., "Increase in the Monozygotic Twinning Rate after Intracytoplasmic Sperm Injection and Blastocyst Stage Embryo Transfer," *Fertility and Sterility* 77, no. 1 (January 1, 2002): 196–98, doi.org/10.1016 /S0015-0282(01)02958-2; Jaime M. Knopman et al., "What Makes Them Split? Identifying Risk Factors That Lead to Monozygotic Twins after In Vitro Fertilization," *Fertility and Sterility* 102, no. 1 (July 1, 2014): 82–89, doi.org/10.1016/j.fertnstert.2014.03.039.

2. Francesco D'Antonio et al., "Perinatal Outcomes of Twin Pregnancies Affected by Early Twin-Twin Transfusion Syndrome: A Systematic Review and Meta-Analysis," *Acta Obstetricia et Gynecologica Scandinavica* 99 (2020): 1121–34, doi.org/10.1111/aogs.13840.

3. Marjolijn S. Spruijt et al., "Twin-Twin Transfusion Syndrome in the Era of Fetoscopic Laser Surgery: Antenatal Management, Neonatal Outcome and Beyond," *Expert Review of Hematology* 13, no. 3 (January 31, 2020): 259–67, doi.org/10.1080/17474086.2020.1720643.

4. Congressional Budget Office, "Mandatory Spending in Fiscal Year 2021: An Infographic," CBO.gov, September 20, 2022, cbo.gov/publication/58270.

5. The Boards of Trustees, Federal Hospital Insurance and Federal Supplementary Medical Insurance Trust Funds, "2022 Annual Report of the Boards of Trustees of the Federal Hospital Insurance and Federal Supplementary

Medical Insurance Trust Funds" (Washington, DC: Centers for Medicare and Medicaid Services, June 2, 2022), cms.gov/files/document/2022-medicare -trustees-report.pdf.

6. Congressional Budget Office, "Health Care," CBO.gov, July 21, 2021, web.archive.org/web/20210701160914/https://www.cbo.gov/topics /health-care.

7. Juliette Cubanski and Tricia Neuman, "What to Know about Medicare Spending and Financing," *KFF* (blog), January 19, 2023, kff.org/medicare /issue-brief/what-to-know-about-medicare-spending-and-financing/.

8. World Health Organization, "Constitution of the World Health Organization," in *Basic Documents*, 49th ed (Geneva: World Health Organization, 2020), 1–19, apps.who.int/iris/handle/10665/339554.

9. Christiaan Monden et al, "Twin Peaks: More Twinning in Humans than Ever Before," *Human Reproduction* 36, no. 6 (2021): 1666–73; I. Blickstein, "Monochorionicity in Perspective," *Ultrasound in Obstetrics & Gynecology* 27, no. 3 (2006): 235–38, doi.org/10.1002/uog.2730.

10. Spruijt et al., "Twin-Twin Transfusion Syndrome in the Era of Fetoscopic Laser Surgery," doi.org/10.1093/humrep/deab029; D'Antonio et al., "Perinatal Outcomes of Twin Pregnancies Affected by Early Twin-Twin Transfusion Syndrome."

11. Centers for Disease Control and Prevention, "Achievements in Public Health, 1900-1999: Healthier Mothers and Babies," 1999, cdc.gov/mmwr/preview /mmwrhtml/mm4838a2.htm.

12. Rachel E. Baker et al, "Infectious Disease in an Era of Global Change | Nature Reviews Microbiology," *Nature Reviews Microbiology* 20 (2022): 193–205, ncbi.nlm.nih.gov/pmc/articles/PMC8513385.

13. "Ten Great Public Health Achievements—United States, 2001–2010," accessed February 2, 2024, cdc.gov/mmwr/preview/mmwrhtml/mm6019a5 .htm.

14. President Biden [@POTUS], "My Father Had an Expression: 'Don't Tell Me What You Value. Show Me Your Budget—and I'll Tell You What You Value.' The Budget I'm Releasing Today Sends a Clear Message That We Value Fiscal Responsibility, Safety and Security, and the Investments Needed to Build a Better America.," Tweet, Twitter, March 28, 2022, twitter.com /POTUS/status/1508530790249947147.

15. USA Spending, "Department of Health and Human Services (HHS) Spending Profile," usaspending.gov, 2022, usaspending.gov/agency/department-of-health-and-human-services; Centers for Disease Control and Prevention, "Office of Financial Resources: FY 2019 Snapshot," 2019.

16. Lauren Weber et al., "Hollowed-Out Public Health System Faces More Cuts Amid Virus," *KFF Health News* (blog), July 1, 2020, khn.org/news/us-public-health-system-underfunded-under-threat-faces-more-cuts-amid-covid-pandemic/.

17. US Census Bureau, "2019 State & Local Government Finance Historical Datasets and Tables," US Census Bureau, 2019, census.gov/data/datasets/2019/econ/local/public-use-datasets.html; Urban Institute, "State and Local Backgrounders: Highway and Road Expenditures," Urban Institute, 2022, urban.org/policy-centers/cross-center-initiatives/state-and-local-finance-initiative/state-and-local-backgrounders/highway-and-road-expenditures; "State and Local Revenues and Expenditures, Per Capita, By Function," Tax Policy Center, July 7, 2023, taxpolicycenter.org/statistics/state-and-local-revenues-and-expenditures-capita-function; "Elementary and Secondary Education Expenditures," Urban Institute, accessed February 2, 2024, urban.org/policy-centers/cross-center-initiatives/state-and-local-finance-initiative/state-and-local-backgrounders/elementary-and-secondary-education-expenditures.

18. Congressional Research Service, "Wildlife Trade, COVID-19, and Other Zoonotic Diseases" (Washington, DC: Congressional Research Service, February 19, 2021), crsreports.congress.gov/product/pdf/IF/IF11494; Jeff Tollefson, "Why Deforestation and Extinctions Make Pandemics More Likely," *Nature* 584, no. 7820 (2020): 175–76, doi.org/10.1038/d41586-020-02341-1.

19. Ève Dubé et al., "Vaccine Hesitancy, Acceptance, and Anti-Vaccination: Trends and Future Prospects for Public Health," *Annual Review of Public Health* 42, no. 1 (2021): 175–91, pubmed.ncbi.nlm.nih.gov/33798403/.

20. G. C. Gray et al., "Adult Adenovirus Infections: Loss of Orphaned Vaccines Precipitates Military Respiratory Disease Epidemics," *Clinical Infectious Diseases: An Official Publication of the Infectious Diseases Society of America* 31, no. 3 (September 2000): 663–70, doi.org/10.1086/313999.

NOTES

CHAPTER 1: PERSEVERANCE

1. World Health Organization, "WHO Organizational Structure," 2022, who .int/about/structure.
2. World Health Organization, "Statue Commemorates Smallpox Eradication," May 17, 2010, who.int/news/item/17-05-2010-statue-commemorates -smallpox-eradication.
3. Milton W. Taylor, "Smallpox," in *Viruses and Man: A History of Interactions* (Cham, Switzerland: Springer International Publishing, 2014), 143–46.
4. Donald R. Hopkins, *The Greatest Killer: Smallpox in History* (Chicago: University of Chicago Press, 2002); Ralph Schram, *A History of the Nigerian Health Services: With an Introd. by Sir Samuel Manuwa* (Ibadan, Nigeria: Ibadan University Press, 1971).
5. Roy Porter, "The Speckled Monster. Smallpox in England, 1670–1970, with Special Reference to Essex," *Medical History* 32, no. 1 (January 1988): 98–99.
6. William Osler, *The Principles and Practice of Medicine: Designed for the Use of Practitioners and Students of Medicine*, 7th ed. (New York: D. Appleton and Company, 1909).
7. Donald A. Henderson, "The Eradication of Smallpox—An Overview of the Past, Present, and Future," *Vaccine* 29, no. S4 (December 30, 2011): D7–9, doi.org/10.1016/j.vaccine.2011.06.080.
8. World Health Organization, "Smallpox," accessed January 10, 2024, who .int/teams/health-product-policy-and-standards/standards-and -specifications/vaccine-standardization/smallpox.
9. Frank Fenner et al., *Smallpox and Its Eradication* (Geneva, Switzerland: World Health Organization, 1988), apps.who.int/iris/handle/10665/39485.
10. Patrick Berche, "Life and Death of Smallpox," *La Presse Médicale, History of Modern Pandemics*, 51, no. 3 (September 1, 2022): 104117, doi.org/10 .1016/j.lpm.2022.104117; Fenner et al., *Smallpox and Its Eradication*.
11. Centers for Disease Control and Prevention, "Clinical Disease: Smallpox," 2016, cdc.gov/smallpox/clinicians/clinical-disease.html; Fenner et al., *Smallpox and Its Eradication*.
12. Emily Anthes, "What the Future May Hold for the Coronavirus and Us," *New York Times*, October 12, 2021, sec. Health, nytimes.com/2021/10/12 /health/coronavirus-mutation-variants.html.
13. W. Seth Carus, "A Short History of Biological Warfare: From Pre-History

to the 21st Century," Occasional Paper (Washington, DC: Center for the Study of Weapons of Mass Destruction, National Defense University Press, August 2017), no. 12, ndupress.ndu.edu/Media/News/Article/1270572/a-short-history-of-biological-warfare-from-pre-history-to-the-21st-century/.

14. Matthew Niederhuber, "The Fight Over Inoculation During the 1721 Boston Smallpox Epidemic," *Science in the News* (blog), December 31, 2014, sitn.hms.harvard.edu/flash/special-edition-on-infectious-disease/2014/the-fight-over-inoculation-during-the-1721-boston-smallpox-epidemic/.

15. Per-Olof Hasselgren, "The Smallpox Epidemics in America in the 1700s and the Role of the Surgeons: Lessons to Be Learned During the Global Outbreak of COVID-19," *World Journal of Surgery* 44, no. 9 (2020): 2837–41, doi.org/10.1007/s00268-020-05670-4; John B. Blake, "Smallpox Inoculation in Colonial Boston," *Journal of the History of Medicine and Allied Sciences* 8, no. 3 (1953): 284–300, doi.org/10.1093/jhmas/VIII.July.284.

16. Blake, "Smallpox Inoculation in Colonial Boston."

17. Donald R. Hopkins, "Smallpox: Ten Years Gone," *American Journal of Public Health* 78, no. 12 (December 1988): 1589–95, doi.org/10.2105/AJPH.78.12.1589; Catherine Thèves, Eric Crubézy, and Philippe Biagini, "History of Smallpox and Its Spread in Human Populations," *Microbiology Spectrum* 4, no. 4 (July 2016), doi.org/10.1128/microbiolspec.poh-0004-2014; Fenner et al., *Smallpox and Its Eradication*.

18. Paul Leicester Ford, *The True George Washington* (Philadelphia, PA: J.B. Lippincott, 1896); Anna Whitelock, *The Queen's Bed: An Intimate History of Elizabeth's Court* (New York: Farrar, Straus and Giroux, 2013).

19. Richard D. Semba, "The Ocular Complications of Smallpox and Smallpox Immunization," *Archives of Ophthalmology* 121, no. 5 (May 2003): 715–19, doi.org/10.1001/archopht.121.5.715.

20. Fenner et al., *Smallpox and Its Eradication*, 231.

21. Stefan Riedel, "Edward Jenner and the History of Smallpox and Vaccination," *Baylor University Medical Center Proceedings* 18, no. 1 (January 2005): 21–25, ncbi.nlm.nih.gov/pmc/articles/PMC1200696/#; Kun Hwang, "Development of Variolation and Its Introduction to Joseon-Era Korea," *Journal of the Korean Society of Traumatology*, 2022, jtraumainj.org/journal/view.php?number=1218.

22. Arthur Boylston, "The Origins of Inoculation," *Journal of the Royal Society of Medicine* 105, no. 7 (July 2012): 309–13, doi.org/10.1258/jrsm.2012.12k044.

23. Riedel, "Edward Jenner and the History of Smallpox and Vaccination"; National Institutes of Health, "Smallpox: A Great and Terrible Scourge," Exhibitions (Bethesda, MD: US National Library of Medicine, July 30, 2013), nlm.nih.gov/exhibition/smallpox/sp_variolation.html.

24. Sophie Hambleton and Ann M. Arvin, "Chickenpox Party or Varicella Vaccine?," in *Hot Topics in Infection and Immunity in Children II*, ed. Andrew J. Pollard and Adam Finn, *Advances in Experimental Medicine and Biology* (Boston, MA: Springer US, 2005), 11–24, doi.org/10.1007/0-387 -25342-4_2.

25. A.J. Tunbridge et al, "Chickenpox in Adults—Clinical Management," *Journal of Infection* 57, no. 2 (August 2008): 95–102, doi.org/10.1016/j.jinf.2008 .03.004.

26. "Pink Book: Varicella," 2022, 022, cdc.gov/vaccines/pubs/pinkbook/vari cella.html.

27. Centers for Disease Control and Prevention, "Chickenpox Vaccination: What Everyone Should Know," April 28, 2021, cdc.gov/vaccines/vpd/var icella/public/index.html; Julie Bosman and Donald G. McNeil Jr., "US Coronavirus Cases are Falling, but Variants Could Erase Progress," *New York Times*, January 22, 2021, 2479962941, nytimes.com/2021/01/22/us /covid-cases-decline.html.

28. The veracity of the milkmaid story is in question. Boylston (2012) finds that Jenner's friend John Baron may have fabricated the tale.

29. Dagny C. Krankowska et al., "Cowpox: How Dangerous Could It Be for Humans? Case Report," *International Journal of Infectious Diseases* 104 (March 2021): 239–41, doi.org/10.1016/j.ijid.2020.12.061.

30. John Hopkins School of Medicine, "Hair of the Cow," Department of the History of Medicine, Johns Hopkins School of Medicine, n.d., hopkinsh istoryofmedicine.org/hair-of-the-cow/.

31. UK National Archives, "Victorian Health Reform: How Did the Victorians View Compulsory Vaccination?," Classroom Resources, The National Archives, accessed October 14, 2022, nationalarchives.gov.uk/education /resources/victorian-health-reform/.

32. Robert M. Wolfe and Lisa K. Sharp, "Anti-Vaccinationists Past and Pres-

ent," *BMJ : British Medical Journal* 325, no. 7361 (August 24, 2002): 430–32, doi.org/10.1136/bmj.325.7361.430.

33. Dorothy Porter and Roy Porter, "The Politics of Prevention: Anti-Vaccinationism and Public Health in Nineteenth-Century England.," *Medical History* 32, no. 3 (July 1988): 231–52, doi.org/10.1017/S00257273000 48225.

34. Local Government Board, England, "The Vaccination Act, 1898," *British Medical Journal* 2, no. 1974 (January 21, 1899): 1351–54, doi.org/10.1136 /bmj.1.1986.176.

35. "When Is a Vaccine Not a Vaccine? Gavi, the Vaccine Alliance," accessed February 2, 2024, gavi.org/vaccineswork/when-vaccine-not-vaccine; The White House, "Press Briefing by Press Secretary Jen Psaki, February 10, 2021," February 10, 2021, whitehouse.gov/briefing-room/press-briefings /2021/02/10/press-briefing-by-press-secretary-jen-psaki-february-10 -2021/.

36. Fenner et al., *Smallpox and Its Eradication*.

37. World Health Organization, "Constitution of the World Health Organization," 1; World Health Organization, "Former Director-General: Dr George Brock Chisholm," April 20, 2017, web.archive.org/web/20170420025405 /http://www.who.int/dg/chisholm/en/.

38. Fenner et al., *Smallpox and Its Eradication*, 366.

39. Fenner et al., *Smallpox and Its Eradication*, 392.

40. Elizabeth Fee et al., "At the Roots of The World Health Organization's Challenges: Politics and Regionalization," *American Journal of Public Health* 106, no. 11 (November 2016): 1912–17, doi.org/10.2105/AJPH.2016.303480; Fenner et al., *Smallpox and Its Eradication*.

41. Fenner et al., *Smallpox and Its Eradication*, 366–67.

42. Fenner et al., *Smallpox and Its Eradication*, 366.

43. B Aylward et al., "When Is a Disease Eradicable? 100 Years of Lessons Learned.," *American Journal of Public Health* 90, no. 10 (October 2000): 1515–20.

44. Fenner et al., *Smallpox and Its Eradication*, 367.

45. Frank Fenner, ed., *Smallpox and Its Eradication*, 198; D. A. Henderson, "Principles and Lessons from the Smallpox Eradication Programme," *Bulletin of the World Health Organization* 65, no. 4 (1987): 535–46.

46. Institute of Medicine, "The Story of Influenza," in *The Threat of Pandemic Influenza: Are We Ready? Workshop Summary* (Washington, DC: National Academies Press, 2005), doi.org/10.17226/11150.

47. World Health Organization, "History of Polio Vaccination," 2021, who .int/news-room/spotlight/history-of-vaccination/history-of-polio -vaccination.

48. Fenner et al., *Smallpox and Its Eradication*, 517, 519.

49. Fenner et al., *Smallpox and Its Eradication*, 785.

50. Fenner et al., *Smallpox and Its Eradication*, 406.

51. J. Donald Millar et al., "Smallpox Vaccination by Intradermal Jet Injection," *Bulletin of the World Health Organization* 41, no. 6 (1969): 749, pubmed .ncbi.nlm.nih.gov/4985446/; National Museum of American History, "Jet Automatic Hypodermic Injection Apparatus—Vaccine Gun," 2011, ameri canhistory.si.edu/collections/nmah_1411247.

52. Joel G. Breman, "Smallpox," *The Journal of Infectious Diseases* 224, no. Suppl 4 (September 30, 2021): S379–86, doi.org/10.1093/infdis/jiaa588.

53. Fenner et al., *Smallpox and Its Eradication*, 473.

54. Andrew W. Artenstein, "Bifurcated Vaccination Needle," *Vaccine* 32, no. 7 (February 7, 2014): 895, doi.org/10.1016/j.vaccine.2013.12.033.

55. Centers for Disease Control and Prevention, "Examples of Major or 'Take' Reactions to Smallpox Vaccination," December 5, 2016, cdc.gov/small pox/clinicians/major-take-reactions-images.html.

56. Fenner et al., *Smallpox and Its Eradication*, 206.

57. J. G. Breman and I. Arita, "The Confirmation and Maintenance of Small-pox Eradication," WHO/SE/80.156 (Geneva, Switzerland: World Health Organization, 1980), nejm.org/doi/full/10.1056/NEJM198011273032204.

58. Donald A. Henderson, *Smallpox: The Death of a Disease: The Inside Story of Eradicating a Worldwide Killer* (Amherst, NY: Prometheus Books, 2009); Fenner et al., *Smallpox and Its Eradication*.

59. Henderson, *Smallpox: The Death of a Disease*, 104.

60. "In Memoriam: Donald Ainslie Henderson MD, MPH '60, 1928–2016," Johns Hopkins Bloomberg School of Public Health, publichealth.jhu.edu /about/history/in-memoriam/donald-a-henderson.

61. Breman and Arita, "The Confirmation and Maintenance of Smallpox Erad-ication," 2.

62. Breman and Arita, "The Confirmation and Maintenance of Smallpox Eradication," 2.

63. Fenner et al., *Smallpox and Its Eradication*, 483.

64. Fenner et al., *Smallpox and Its Eradication*, 504.

65. Fenner et al., *Smallpox and Its Eradication*, 310.

66. Fenner et al., *Smallpox and Its Eradication*, 764.

67. Fenner et al., *Smallpox and Its Eradication*, 1021.

68. Breman and Arita, "The Confirmation and Maintenance of Smallpox Eradication," 15.

69. Fenner, et al., *Smallpox and Its Eradication*, 533.

70. Amy McDermott, "Herd Immunity Is an Important—and Often Misunderstood—Public Health Phenomenon," *Proceedings of the National Academy of Sciences* 118, no. 21 (May 25, 2021): e2107692118, doi.org/10.1073/pnas.2107692118.

71. David H. Molyneux et al., "Disease Eradication, Elimination and Control: The Need for Accurate and Consistent Usage," *Trends in Parasitology* 20, no. 8 (August 1, 2004): 347–51, doi.org/10.1016/j.pt.2004.06.004.

72. Centers for Disease Control and Prevention, "Measles Cases and Outbreaks," January 26, 2024, cdc.gov/measles/cases-outbreaks.html.

73. Molyneux et al., "Disease Eradication, Elimination and Control."

74. Fenner et al., *Smallpox and Its Eradication*, 366–67.

75. Institute of Medicine, *The Future of Public Health* (Washington, DC: National Academies Press, 1988), doi.org/10.17226/1091.

76. Anna M. Acosta et al., "Diphtheria," in *Epidemiology and Prevention of Vaccine-Preventable Diseases*, ed. Elisha Hall et al., 14th ed. (Washington, DC: Public Health Foundation, 2021), 97–110.

77. A Rick Cooper, "Diphtheria," *Paediatrics & Child Health* 7, no. 3 (March 2002): 205; Centers for Disease Control and Prevention, "Clinical Information | CDC," September 11, 2023, cdc.gov/diphtheria/clinicians.html.

78. Rob Roy MacGregor, "206—Corynebacterium Diphtheriae (Diphtheria)," in *Mandell, Douglas, and Bennett's Principles and Practice of Infectious Diseases* (Eighth Edition), ed. John E. Bennett, Raphael Dolin, and Martin J. Blaser (Philadelphia: W.B. Saunders, 2015), 2366-2372.e1, doi

.org/10.1016/C2012-1-00075-6; Curtis Welch, "The Diphtheria Epidemic at Nome," *Journal of the American Medical Association* 84, no. 17 (April 25, 1925): 1290–91, doi.org/10.1001/jama.1925.02660430048031.

79. E. D. Stokes, "The Race for Life," *Public Health Reports* 111, no. 3 (June 1996): 272–75, stacks.cdc.gov/view/cdc/64371.

80. Smithsonian Institution, "Alaskan Dog Sled Mail Carrier," *Smithsonian Snapshot* (blog), January 10, 2012, si.edu/newsdesk/snapshot/alaskan-dog -sled-mail-carrier.

81. Stokes, "The Race for Life."

82. Stokes, "The Race for Life."

83. Stokes, "The Race for Life."

84. Richard Raponi, "Balto vs. the Alaskan Black Death," Cleveland Historical, accessed October 14, 2022, clevelandhistorical.org/items/show/610.

85. Stokes, "The Race for Life."

86. Steve Moyer, "Dog vs. Airplane," *Humanities*, December 2013, neh.gov /humanities/2013/novemberdecember/feature/dog-vs-airplane.

87. Alexandra Lord and Diane Wendt, "How Vaccines, a Collective Triumph of Modern Medicine, Conquered the World's Diseases," *Smithsonian Magazine*, September 28, 2015, smithsonianmag.com/smithsonian-institution /how-cow-dogs-doctors-delivered-some-greatests-triumphs-modern -medicine-180956672/.

88. Von E. Martin, *A Long Way To Nome: The Serum Run '25 Expedition*, First Edition (CreateSpace Independent Publishing Platform, 2009).

89. Gay Salisbury and Laney Salisbury, *The Cruelest Miles: The Heroic Story of Dogs and Men in a Race Against an Epidemic* (W. W. Norton & Company, 2003).

90. Don Bowers, "Iditarod Race History," Iditarod, May 13, 2020, iditarod .com/race-history/.

91. Fenner et al., *Smallpox and Its Eradication*, 395, 540, 1097.

92. Global Polio Eradication Initiative, "Remembering Ali Maalin," September 26, 2018, polioeradication.org/news-post/remembering-ali-maalin/.

93. Global Polio Eradication Initiative, "Remembering Ali Maalin."

94. Global Polio Eradication Initiative, "Remembering Ali Maalin."

95. R. A. Shooter et al., "Report of the Investigation into the Cause of the 1978 Birmingham Smallpox Occurrence," Commissioned by the House of

Commons (London, United Kingdom, 1980), 51, assets.publishing.service
.gov.uk/government/uploads/system/uploads/attachment_data/file
/228654/0668.pdf.pdf.

96. R. A. Shooter et al., "Report of the Investigation into the Cause of the
1978 Birmingham Smallpox Occurrence," 186–91.

97. Adamson S. Muula, "Does the 'World' Still Need to Keep Live Samples of
the Smallpox Virus?," *Malawi Medical Journal* 34, no. 2 (June 2022): 72,
doi.org/10.4314/mmj.v34i2.1.

98. Caroline Schuerger et al., "Mapping Biosafety Level-3 Laboratories by
Publications" (Washington, DC: Center for Security and Emerging Technology, August 2022), doi.org/10.51593/20220019.

CHAPTER 2: COMPLACENCY

1. D. H. Rosenblatt, "Unexploded Ordnance Issues at Aberdeen Proving
Ground: Background Information," Report (Argonne, IL: Argonne National Laboratory, November 1, 1996), doi.org/10.2172/437674.

2. Lauren Nelson, "100 Years of Excellence: The ATC Story, Part 3," US
Army Aberdeen Test Center, accessed December 28, 2023, https://www
.atec.army.mil/atc/ATC_PointPosition/October_Vol1Num3/ATC_His
tory_Part3.html.

3. Ian R. Tizard and Jeffrey M. B. Musser, "Panic and Neglect—2000–2018,"
Great American Diseases, 2022, 319–44, doi.org/10.1016/B978-0-323-98925-1
.00004-0.

4. Maria Gilson deValpine, "Influenza in Bristol Bay, 1919: 'The Saddest Re-
pudiation of a Benevolent Intention,'" *SAGE Open* 5, no. 1 (March 1,
2015), doi.org/10.1177%2F2158244015577418; Sandra M. Tomkins, "The
Influenza Epidemic of 1918–19 in Western Samoa," *The Journal of Pacific
History* 27, no. 2 (1992): 181–97, jstor.org/stable/25169127.

5. Nancy Tomes, "'Destroyer and Teacher': Managing the Masses During
the 1918–1919 Influenza Pandemic," *Public Health Reports* 125, no. Suppl
3 (2010): 48–62, doi.org/10.1177/00333549101250S308.

6. John Graham Royde-Smith, ed., "World War II—Costs of the War," in
Encyclopedia Britannica, accessed July 3, 2022, britannica.com/event/World
-War-II/Costs-of-the-war; Terrence M. Tumpey et al., "Characterization

of the Reconstructed 1918 Spanish Influenza Pandemic Virus," *Science* 310, no. 5745 (October 7, 2005): 77–80, doi.org/10.1126/science.1119392.

7. Centers for Disease Control and Prevention, "1918 Pandemic Influenza Historic Timeline," March 20, 2018, stacks.cdc.gov/view/cdc/119435.

8. Jeffery K. Taubenberger and David M. Morens, "1918 Influenza: The Mother of All Pandemics," *Emerging Infectious Diseases* 12, no. 1 (January 2006): 15–22, doi.org/10.3201/eid1201.050979.

9. N. R. Grist, "Pandemic Influenza 1918," *British Medical Journal* 2, no. 6205 (December 22, 1979): 1632, doi.org/10.1136/bmj.2.6205.1632.

10. Grist, "Pandemic Influenza 1918," 1632.

11. Carol R. Byerly, "The US Military and the Influenza Pandemic of 1918–1919," *Public Health Reports* 125, no. Suppl 3 (2010): 82–91. *Public Health Reports* 125, no. S3 (2010): 82–91, ncbi.nlm.nih.gov/pmc/articles/PMC2862337/.

12. Byerly, "The US Military and the Influenza Pandemic of 1918–1919," 82.

13. Centers for Disease Control and Prevention, "1918 Pandemic Influenza Historic Timeline."

14. Merrite W. Ireland, "Excerpts on the Influenza and Pneumonia Pandemic of 1918," Surgeon General, US Army, 1919, achh.army.mil/history/book -wwi-1918flu-arsg1919-arsg1919intro.

15. Robert J. Smith, "The U.S. Army and the Great Influenza Pandemic of 1918," *On Point* 25, no. 2 (2019), history.army.mil/covid19/Spanish-Flu1918 _On-Point-Magazine-Dr-Bob-Smith.pdf.

16. Byerly, "The U.S. Military and the Influenza Pandemic of 1918–1919," US Surgeon-General, *The Medical Department of the U.S. Army in the World War* (Washington, DC: US Government Printing Office, 1926).

17. *The Medical Department of the U.S. Army in the World War.*

18. Smith, "The U.S. Army and the Great Influenza Pandemic of 1918," 23.

19. Byerly, "The U.S. Military and the Influenza Pandemic of 1918–1919."

20. Máire A. Connolly and David L. Heymann, "Deadly Comrades: War and Infectious Diseases," *The Lancet* 360 (December 2002): s23, doi.org/10.1016 /S0140-6736(02)11807-1.

21. Smith, "The U.S. Army and the Great Influenza Pandemic of 1918," 25.

22. Willard J. Stone and George W. Swift, "Influenza and Influenzal Pneumonia at Fort Riley, Kansas: From Sept. 16 to Nov. 1, 1918," *Journal of the*

American Medical Association 72, no. 7 (February 15, 1919): 487, doi.org
/10.1001/jama.1919.02610070025014.

23. Byerly, "The U.S. Military and the Influenza Pandemic of 1918–1919."

24. Theodore E. Woodward, *Armed Forces Epidemiological Board: Its First Fifty Years, 1940–1990*, ed. Russ Zajtchuk et al. (Falls Church, Virginia: Office of the Surgeon General, Department of the Army, 1990), Top.

25. Sanders Marble, "Brigadier General James Stevens Simmons (1890–1954), Medical Corps, United States Army: A Career in Preventive Medicine," *Journal of Medical Biography* 20, no. 1 (February 2012): 3–10, doi.org /10.1258/jmb.2010.010054.

26. "Naval Health Research Center Investigators Track Vaccine in Military Recruits," Navy Medicine, accessed January 28, 2024, https://www.med .navy.mil/Media/News/Article/2608788/naval-health-research-center -investigators-track-vaccine-in-military-recruits/.

27. M. R. Hilleman et al., "Epidemiologic Investigations with Respiratory Disease Virus RI-67," *American Journal of Public Health* 45, no. 2 (February 1955): 203–10, ajph.aphapublications.org/doi/pdf/10.2105/AJPH.45.2.203.

28. Franklin H. Top, "Control of Adenovirus Acute Respiratory Disease in US Army Trainees," *The Yale Journal of Biology and Medicine* 48, no. 3 (July 1975): 185–95, ncbi.nlm.nih.gov/pmc/articles/PMC2595226/.

29. Niranjan Balliram, "Adenovirus Vaccine Shortfall: Impact on Readiness and Deployability," Strategy Research Project (Carlisle Barracks, PA: US Army War College, December 13, 2001), apps.dtic.mil/sti/citations/ADA403017.

30. Sean McBride, "Army Corps Repairs Aging Reception Battalion Barracks on Fort Jackson," US Army Corps of Engineers, Charleston Division, August 25, 2020, https://www.sac.usace.army.mil/Media/News-Stories/Article/2324107 /army-corps-repairs-aging-reception-battalion-barracks-on-fort-jackson/.

31. Department of the Army, "Uniform and Insignia: Wear and Appearance of Army Uniforms and Insignia," Army Regulation 670-1 (Washington, DC: Department of the Army, January 26, 2021), armypubs.army.mil /epubs/DR_pubs/DR_a/ARN30302-AR_670-1-000-WEB-1.pdf.

32. Department of the Army, "Special Forces Use of Pack Animals," Field Manual 3-05.213 (Washington, DC: Department of the Army, June 2004), 10–2, irp.fas.org/doddir/army/fm3-05-213.pdf.

33. Department of the Army, "Army Facilities Management," Army Regulation 420-1 (Washington, DC: Department of the Army, August 24, 2012), 140.

34. US Army, "Facility Sanitation Controls and Inspections" (Washington, DC: US Army, March 1, 2019).

35. Department of the Army, "Enlisted Initial Entry Training Policies and Administration," TRADOC Regulation 350-6 (Washington, DC: Department of the Army, December 8, 2022), 47.

36. Caroline Brodkey and Joel C. Gaydos, "United States Army Guidelines for Troop Living Space: A Historical Review," *Military Medicine* 145, no. 6 (June 1980): 418, academic.oup.com/milmed/article-abstract/145/6/418 /4904394.

37. Brodkey and Gaydos, "United States Army Guidelines for Troop Living Space: A Historical Review," 418.

38. Katherine E. Randall et al., "How Did We Get Here: What Are Droplets and Aerosols and How Far Do They Go? A Historical Perspective on the Transmission of Respiratory Infectious Diseases," *Interface Focus* 11, no. 6 (October 12, 2021), doi.org/10.1098/rsfs.2021.0049.

39. Hilleman et al., "Epidemiologic Investigations with Respiratory Disease Virus RI-67."

40. Maurice R. Hilleman, "Personal Historical Chronicle of Six Decades of Basic and Applied Research in Virology, Immunology, and Vaccinology," *Immunological Reviews* 170, no. 1 (August 1999): 7–27, doi.org/10.1111 /j.1600-065X.1999.tb01325.x.

41. Richard Conniff, "A Forgotten Pioneer of Vaccines," *New York Times*, May 6, 2013, sec. D, nytimes.com/2013/05/07/health/maurice-hilleman-mmr -vaccines-forgotten-hero.html.

42. Hilleman, "Personal Historical Chronicle of Six Decades of Basic and Applied Research in Virology, Immunology, and Vaccinology."

43. Hilleman, "Personal Historical Chronicle of Six Decades of Basic and Applied Research in Virology, Immunology, and Vaccinology."

44. Hilleman, "Personal Historical Chronicle of Six Decades of Basic and Applied Research in Virology, Immunology, and Vaccinology."

45. Harold S. Ginsberg, ed., *The Adenoviruses*, The Viruses (New York, NY: Plenum Press, 1984), doi.org/10.1007/978-1-4684-7935-5.

46. Wallace P. Rowe et al., "Isolation of a Cytopathogenic Agent from Human

Adenoids Undergoing Spontaneous Degeneration in Tissue Culture," *Proceedings of the Society for Experimental Biology and Medicine* 84, no. 3 (December 1953): 570–73, doi.org/10.3181/00379727-84-20714.

47. Christopher M. Robinson et al., "Molecular Evolution of Human Species D Adenoviruses," *Infection, Genetics and Evolution* 11, no. 6 (August 2011): 1208–17, doi.org/10.1016/j.meegid.2011.04.031.

48. Hilleman et al., "Epidemiologic Investigations with Respiratory Disease Virus RI-67," 207.

49. Subrat Khanal et al, "The Repertoire of Adenovirus in Human Disease: The Innocuous to the Deadly," *Biomedicines* 6, no. 1 (March 7, 2018): 30, doi.org/10.3390/biomedicines6010030.

50. Khanal et al "The Repertoire of Adenovirus in Human Disease: The Innocuous to the Deadly."

51. Khanal et al "The Repertoire of Adenovirus in Human Disease: The Innocuous to the Deadly."

52. Robert N. Potter et al., "Adenovirus-Associated Deaths in US Military during Postvaccination Period, 1999–2010," *Emerging Infectious Diseases* 18, no. 3 (March 2012): 507–9, doi.org/10.3201/eid1803.111238.

53. Centers for Disease Control and Prevention, "Adenoviruses: Outbreaks," February 17, 2023, cdc.gov/adenovirus/outbreaks.html.

54. Charles H. Hoke Jr. and Clifford E. Snyder Jr., "History of the Restoration of Adenovirus Type 4 and Type 7 Vaccine, Live Oral (Adenovirus Vaccine) in the Context of the Department of Defense Acquisition System," *Vaccine* 31, no. 12 (March 15, 2013): 1623–32, doi.org/10.1016/j.vaccine.2012.12.029.

55. M. R. Hilleman et al., "Prevention of Acute Respiratory Illness in Recruits by Adenovirus (RI-APC-ARD) Vaccine," *Proceedings of the Society for Experimental Biology and Medicine* 92, no. 2 (June 1956): 377–83, doi.org/10.3181/00379727-92-22484.

56. Paul T. Arlt, "'On Target!'—New York Herald Tribune Cartoon Regarding Dr. Maurice Hilleman's Adenovirus Vaccine," 1956, glossy print, 10 in × 8 in, 1956, 2017.3081.07, National Museum of American History, americanhistory.si.edu/collections/nmah_1876996.

57. Hoke and Snyder, "History of the Restoration of Adenovirus Type 4 and Type 7 Vaccine, Live Oral (Adenovirus Vaccine) in the Context of the Department of Defense Acquisition System," March 15, 2013.

58. Balliram, "Adenovirus Vaccine Shortfall."

59. Peter B. Collis et al., "Adenovirus Vaccines in Military Recruit Populations: A Cost-Benefit Analysis," *The Journal of Infectious Diseases* 128, no. 6 (December 1973): 745–52, doi.org/10.1093/infdis/128.6.745.

60. Jose L. Sanchez et al., "Respiratory Infections in the U.S. Military: Recent Experience and Control," *Clinical Microbiology Reviews* 28, no. 3 (June 17, 2015): 743–800, doi.org/10.1128/cmr.00039-14.

61. G. C. Gray et al., "Adult Adenovirus Infections: Loss of Orphaned Vaccines Precipitates Military Respiratory Disease Epidemics," *Clinical Infectious Diseases: An Official Publication of the Infectious Diseases Society of America* 31, no. 3 (September 2000): 663–70, doi.org/10.1086/313999; Hoke and Snyder, "History of the Restoration of Adenovirus Type 4 and Type 7 Vaccine, Live Oral (Adenovirus Vaccine) in the Context of the Department of Defense Acquisition System," March 15, 2013.

62. Balliram, "Adenovirus Vaccine Shortfall," 7.

63. Office of Management and Budget, Executive Office of the President, "Budget FY 2021—Historical Tables, Budget of the United States Government, Fiscal Year 2021," Office of Management and Budget, Executive Office of the President, February 10, 2020, govinfo.gov/app/details/BUDGET-2021-TAB/summary.

64. Based on personal correspondence with military public health officials involved with the adenovirus program at the time.

65. Charles H. Hoke and Clifford E. Snyder, "History of the Restoration of Adenovirus Type 4 and Type 7 Vaccine, Live Oral (Adenovirus Vaccine) in the Context of the Department of Defense Acquisition System," 1623–32.

66. Personal correspondence, see note 64.

67. Personal correspondence, see note 64.

68. Personal correspondence, see note 64.

69. Sarah Lueck, "Boot-Camp Bug Returns to the Barracks When Pentagon Pulls the Plug on Vaccine," *Wall Street Journal*, July 13, 2001, sec. Front Section, wsj.com/articles/SB994971407564921950.

70. Balliram, "Adenovirus Vaccine Shortfall," 9.

71. Angela E. Micah et al., "Global Investments in Pandemic Preparedness and COVID-19: Development Assistance and Domestic Spending on Health between 1990 and 2026," *The Lancet Global Health* 11, no. 3 (March 1, 2023): e385–413, doi.org/10.1016/S2214-109X(23)00007-4.

72. Stanley M. Lemon et al., eds., "Appendix A. Urgent Attention Needed to Restore Lapsed Adenovirus Vaccine Availability," in *Protecting Our Forces: Improving Vaccine Acquisition and Availability in the U.S. Military,* Institute of Medicine (Washington, DC: National Academies Press, 2002), ncbi .nlm.nih.gov/books/NBK220956/.

73. Centers for Disease Control and Prevention, "Two Fatal Cases of Adenovirus-Related Illness in Previously Healthy Young Adults—Illinois, 2000," *Morbidity and Mortality Weekly Report* 50, no. 26 (July 6, 2001): 553–55; Kevin L. Russell et al., "Transmission Dynamics and Prospective Environmental Sampling of Adenovirus in a Military Recruit Setting," *The Journal of Infectious Diseases* 194, no. 7 (October 1, 2006): 877–85, doi.org/10.1086 /507426.

74. Gray et al., "Adult Adenovirus Infections."

75. Gray et al., "Adult Adenovirus Infections"; M. René Howell et al., "Prevention of Adenoviral Acute Respiratory Disease in Army Recruits: Cost-Effectiveness of a Military Vaccination Policy," *American Journal of Preventive Medicine* 14, no. 3 (April 1, 1998): 168–75, doi.org/10.1016/S0749-3797(97) 00064-0.

76. Centers for Disease Control and Prevention, "Two Fatal Cases of Adenovirus-Related Illness in Previously Healthy Young Adults—Illinois, 2000."

77. Lueck, "Boot-Camp Bug Returns to the Barracks When Pentagon Pulls the Plug on Vaccine."

78. Centers for Disease Control and Prevention, "Two Fatal Cases of Adenovirus-Related Illness in Previously Healthy Young Adults—Illinois, 2000."

79. Lemon et al., "Appendix A. Urgent Attention Needed to Restore Lapsed Adenovirus Vaccine Availability."

80. Michael J. Berens, "'Major Screw-up': Boot-Camp Virus Runs Rampant," *The Seattle Times,* October 3, 2004, archive.seattletimes.com/archive/?date= 20041003&slug=virus03m.

81. Hoke and Snyder, "History of the Restoration of Adenovirus Type 4 and Type 7 Vaccine, Live Oral (Adenovirus Vaccine) in the Context of the Department of Defense Acquisition System," March 2013.

82. Berens, "'Major Screw-up': Boot-Camp Virus Runs Rampant."

83. Hoke and Snyder, "History of the Restoration of Adenovirus Type 4 and Type 7 Vaccine, Live Oral (Adenovirus Vaccine) in the Context of the Department of Defense Acquisition System," March 2013.

84. Potter et al., "Adenovirus-Associated Deaths in US Military during Post-vaccination Period, 1999–2010," March 2012, 507.

85. Potter et al., "Adenovirus-Associated Deaths in US Military during Post-vaccination Period, 1999–2010," 507–9.

CHAPTER 3: SKILLS

1. Michael A. E. Ramsay, "John Snow, MD: Anaesthetist to the Queen of England and Pioneer Epidemiologist," *Baylor University Medical Center Proceedings* 19, no. 1 (January 2006): 24, ncbi.nlm.nih.gov/pmc/articles /PMC1325279/.

2. Ramsay, "John Snow, MD."

3. Dominique Buchillet, "Epidemic Diseases in the Past: History, Philosophy, and Religious Thought," in *Encyclopedia of Infectious Diseases: Modern Methodologies*, ed. Michael Tibayrenc (Hoboken, NJ: John Wiley & Sons, Inc., 2007), 522, horizon.documentation.ird.fr/exl-doc/pleins_textes/divers19 -11/010041106.pdf.

4. World Health Organization, "Cholera," accessed January 1, 2024, who .int/health-topics/cholera.

5. Stephen Halliday, "Death and Miasma in Victorian London: An Obstinate Belief," *BMJ* 323, no. 7327 (December 22, 2001): 1469–71, doi.org /10.1136/bmj.323.7327.1469.

6. Theodore H. Tulchinsky, "John Snow, Cholera, the Broad Street Pump; Waterborne Diseases Then and Now," *Case Studies in Public Health*, 2018, 77–99, doi.org/10.1016/B978-0-12-804571-8.00017-2.

7. John Snow, *On the Mode of Transmission of Cholera*, 2nd ed. (London: Churchill, 1855) reproduced in W. H. Frost, ed., *Snow on Cholera: A Reprint of Two Papers* by John Snow, M.D. (New York: The Commonwealth Fund, 1936), 38–56.

8. John Snow, "'Dr. Snow's Report,' in the Report on the Cholera Outbreak in the Parish of St. James, Westminster, during the Autumn of 1854," Cholera Inquiry Committee, July 1855, The John Snow Archive and Research Companion, johnsnow.matrix.msu.edu/work.php?id=15-78-55.

9. John M. Eyler, "William Farr on the Cholera: The Sanitarian's Disease Theory and the Statistician's Method," *Journal of the History of Medicine*

and Allied Sciences XXVIII, no. 2 (April 1973): 79–100, doi.org/10.1093 /jhmas/XXVIII.2.79.

10. Paul Bingham et al, "John Snow, William Farr and the 1849 Outbreak of Cholera That Affected London: A Reworking of the Data Highlights the Importance of the Water Supply," *Public Health* 118, no. 6 (September 1, 2004): 387–94, doi.org/10.1016/j.puhe.2004.05.007.

11. Ray M. Merrill and Thomas C. Timmreck, *Introduction to Epidemiology*, 4th ed (Sudbury, MA: Jones and Bartlett Publishers, 2006).

12. Alison Stargel et al., "Case Investigation and Contact Tracing Efforts from Health Departments in the United States, November 2020 to December 2021," *Clinical Infectious Diseases* 75, no. S2 (October 1, 2022): S326–33, doi.org/10.1093/cid/ciac442.

13. Yonatan H. Grad and Marc Lipsitch, "Epidemiologic Data and Pathogen Genome Sequences: A Powerful Synergy for Public Health," *Genome Biology* 15, no. 11 (November 18, 2014): 538, doi.org/10.1186/s13059-014 -0538-4.

14. Michael M. Wagner et al., "Design of a National Retail Data Monitor for Public Health Surveillance," *Journal of the American Medical Informatics Association* 10, no. 5 (September 2003): 409–18, doi.org/10.1197/jamia .M1357; Shaoyang Ning et al., "Accurate Regional Influenza Epidemics Tracking Using Internet Search Data," *Scientific Reports* 9 (March 27, 2019): 5238, doi.org/10.1038/s41598-019-41559-6; Jennifer M. Radin et al., "The Hopes and Hazards of Using Personal Health Technologies in the Diagnosis and Prognosis of Infections," *The Lancet Digital Health* 3, no. 7 (July 2021): e455-61, doi.org/10.1016/S2589-7500(21)00064-9.

15. Jay E. Gee et al., "Multistate Outbreak of Melioidosis Associated with Imported Aromatherapy Spray," *New England Journal of Medicine* 386, no. 9 (March 3, 2022): 861–68, doi.org/10.1056/NEJMoa2116130.

16. Centers for Disease Control and Prevention, "What Increases Your Risk," April 30, 2024, cdc.gov/melioidosis/prevention/#cdc_prevention_risk-what -increases-your-risk.

17. Centers for Disease Control and Prevention, "Melioidosis and Cases Around the World," May 1, 2024, cdc.gov/melioidosis/risk-factors/.

18. Tulchinsky, "John Snow, Cholera, the Broad Street Pump," 77.

19. Frank Ching, "Bird Flu, SARS and Beyond," *130 Years of Medicine in Hong*

Kong, March 15, 2018, 381–434, doi.org/10.1007/978-981-10-6316-9_14; Ellen Nakashima, "SARS Signals Missed in Hong Kong," *Washington Post*, May 20, 2003, washingtonpost.com/archive/politics/2003/05/20/sars-signals -missed-in-hong-kong/50ff4807-4862-4229-8bbd-ec5932b5c896/.

20. Ching, "Bird Flu, SARS and Beyond"; Nakashima, "SARS Signals Missed in Hong Kong."

21. Ching, "Bird Flu, SARS and Beyond"; Nakashima, "SARS Signals Missed in Hong Kong."

22. Ching, "Bird Flu, SARS and Beyond."

23. Nakashima, "SARS Signals Missed in Hong Kong."

24. WHO Western Pacific Region, *SARS: How a Global Epidemic Was Stopped* (Manila, Philippines: World Health Organization Regional Office for the Western Pacific, 2006), apps.who.int/iris/handle/10665/207501.

25. Margot Cohen, Peter Fritsch, and Matt Pottinger, "Blue-Jean Maker Johnny Chen Became Host of Mystery Illness," *Wall Street Journal*, March 19, 2003, wsj.com/articles/SB104802055563183300.

26. Ellen Nakashima, "Vietnam Took Lead in Containing SARS," *Washington Post*, May 5, 2003, washingtonpost.com/archive/politics/2003/05/05 /vietnam-took-lead-in-containing-sars/b9b97e91-b325-42f9-98ef-e23 da9f257a0/.

27. Cohen, Fritsch, and Pottinger, "Blue-Jean Maker Johnny Chen Became Host of Mystery Illness."

28. Walter N. Harrington et al., "The Evolution and Future of Influenza Pandemic Preparedness," *Experimental & Molecular Medicine* 53, no. 5 (May 2021): 737–49, doi.org/10.1038/s12276-021-00603-0; Jeffery K. Taubenberger and David M. Morens, "Influenza: The Once and Future Pandemic," *Public Health Reports* 125, no. Suppl 3 (2010): 16–26.

29. Ivan Oransky, "Carlo Urbani," *The Lancet* 361 (April 26, 2003): 1481, thelancet.com/journals/lancet/article/PIIS0140-6736(03)13107-8/full text.

30. Donald G. McNeil, Jr., "Disease's Pioneer Is Mourned as a Victim," *New York Times*, April 8, 2003, nytimes.com/2003/04/08/science/disease-s-pioneer -is-mourned-as-a-victim.html.

31. WHO Western Pacific Region, *SARS: How a Global Epidemic Was Stopped*, 95.

32. Nakashima, "Vietnam Took Lead in Containing SARS."

33. World Health Organization, "Update 31—Coronavirus Never before Seen in Humans Is the Cause of SARS," News Release, April 16, 2003, who.int /news/item/16-04-2003-update-31---coronavirus-never-before -seen-in-humans-is-the-cause-of-sars.

34. James D. Cherry and Paul Krogstad, "SARS: The First Pandemic of the 21st Century," *Pediatric Research* 56, no. 1 (2004): 1–5, doi.org/10.1203 /01.PDR.0000129184.87042.FC.

35. SARS-CoV is the virus that causes the disease SARS, like HIV is the virus that causes AIDS.

36. World Health Organization, "World Health Organization Issues Emergency Travel Advisory," March 15, 2003, who.int/news/item/15-03-2003 -world-health-organization-issues-emergency-travel-advisory.

37. Oransky, "Carlo Urbani."

38. Centers for Disease Control and Prevention, "CDC Support to Health Officials: How CDC Can Help Respond to Emerging Local Health Concerns," March 3, 2023, web.archive.org/web/20220525142433/https://www .cdc.gov/publichealthgateway/healthdepartmentresources/health-official -support.html.

39. Renee Ghert-Zand, "New Study Shows Israel Experiencing Unprecedented Outbreak of 'Fifth Disease,'" accessed February 2, 2024, timesofisrael.com /new-study-shows-israel-experiencing-unprecedented-outbreak-of-fifth -disease/.

40. Nakashima, "SARS Signals Missed in Hong Kong."

41. Institute of Medicine Forum on Microbial Threats, *Learning from SARS: Preparing for the Next Disease Outbreak: Workshop Summary*, ed. Stacey Knobler et al. (Washington, DC: National Academies Press, 2004), ncbi .nlm.nih.gov/books/NBK92462/.

42. Institute of Medicine Forum on Microbial Threats. Donald E. Low, "SARS: Lessons from Toronto," in *Learning from SARS*.

43. World Health Organization, "Consensus Document on the Epidemiology of Severe Acute Respiratory Syndrome (SARS)," WHO/CDS/CSR/GAR/ 2003.11 (Geneva, Switzerland: World Health Organization, May 17, 2003), who.int/publications/i/item/consensus-document-on-the-epidemiology-of -severe-acute-respiratory-syndrome-(-sars).

44. Thomas Abraham, *Twenty-First Century Plague: The Story of SARS* (Baltimore, MD: Johns Hopkins University Press, n.d.); Hoe Nam Leong and Hong Huay Lim, "SARS—My Personal Battle," *Travel Medicine and Infectious Disease* 9, no. 3 (May 2011): 109–12, doi.org/10.1016/j.tmaid.2010.10.007.

45. Leong and Lim, "SARS—My Personal Battle," 110.

46. Leong and Lim, "SARS—My Personal Battle," 110.

47. WHO Western Pacific Region, *SARS: How a Global Epidemic Was Stopped*.

48. Yousef Alimohamadi et al., "Case Fatality Rate of COVID-19: A Systematic Review and Meta-Analysis," *Journal of Preventive Medicine and Hygiene* 62, no. 2 (July 30, 2021): E311–20, doi.org/10.15167/2421-4248/jpmh2021.62.2.1627; estimates for case fatality risk vary widely by geography, age group, and methodology.

49. World Health Organization, "Consensus Document on the Epidemiology of Severe Acute Respiratory Syndrome (SARS)."

50. Shriyansh Srivastava et al., "Emergence of Marburg Virus: A Global Perspective on Fatal Outbreaks and Clinical Challenges," *Frontiers in Microbiology* 14 (September 13, 2023), doi.org/10.3389/fmicb.2023.1239079.

51. Lawrence K. Altman, "In Philadelphia 30 Years Ago, an Eruption of Illness and Fear," *New York Times*, August 1, 2006, nytimes.com/2006/08/01/health/01docs.html.

52. Altman, "In Philadelphia 30 Years Ago, an Eruption of Illness and Fear."

53. Elana Gordon, "40 Years Later, Scientist Who First Discovered Legionnaires' Disease Is Still Learning Lessons," *The Pulse* (WHYY, July 28, 2016), 3:33, whyy.org/articles/40-years-later-scientist-who-first-discovered-legionnaires-disease-is-still-learning-lessons/.

54. Gordon, "40 Years Later, Scientist Who First Discovered Legionnaires' Disease Is Still Learning Lessons."

55. World Health Organization, "Legionellosis," September 6, 2022, who.int/news-room/fact-sheets/detail/legionellosis.

56. National Institute of Allergy and Infectious Diseases, "Coronaviruses," National Institutes of Health (NIH), March 22, 2022, https://www.niaid.nih.gov/diseases-conditions/coronaviruses.

57. World Health Organization, "Summary of Probable SARS Cases with Onset of Illness from 1 November 2002 to 31 July 2003," July 24, 2015,

who.int/publications/m/item/summary-of-probable-sars-cases-with-onset-of-illness-from-1-november-2002-to-31-july-2003.

58. WHO Western Pacific Region, *SARS: How a Global Epidemic Was Stopped.*

59. Don Klinkenberg et al., "The Effectiveness of Contact Tracing in Emerging Epidemics," *PLOS ONE* 1, no. 1 (December 20, 2006): e12, doi.org/10.1371/journal.pone.0000012.

60. Shrivastava Saurabh and Shrivastava Prateek, "Role of Contact Tracing in Containing the 2014 Ebola Outbreak: A Review," *African Health Sciences* 17, no. 1 (May 2017): 225–36, doi.org/10.4314/ahs.v17i1.28; Wafaa M. El-Sadr et al., "Contact Tracing: Barriers and Facilitators," *American Journal of Public Health* 112, no. 7 (July 2022): 1025–33, doi.org/10.2105/AJPH.2022.306842.

61. Tadele Girum et al., "Global Strategies and Effectiveness for COVID-19 Prevention through Contact Tracing, Screening, Quarantine, and Isolation: A Systematic Review," *Tropical Medicine and Health* 48, no. 1 (November 23, 2020): 91, doi.org/10.1186/s41182-020-00285-w.

62. City of New York, "Transcript: Mayor de Blasio Holds Media Availability on COVID-19," The official website of the City of New York, April 16, 2020, nyc.gov/office-of-the-mayor/news/260-20/transcript-mayor-de-blasio-holds-media-availability-covid-19.

63. Gregg Colburn et al., "Hotels as Noncongregate Emergency Shelters: An Analysis of Investments in Hotels as Emergency Shelter in King County, Washington During the COVID-19 Pandemic," *Housing Policy Debate* 32, no. 6 (November 2, 2022): 853–75, doi.org/10.1080/10511482.2022.2075027.

64. World Health Organization, "Consensus Document on the Epidemiology of Severe Acute Respiratory Syndrome (SARS)."

65. Severe Acute Respiratory Syndrome Epidemiology Working Group, "Consensus Document on the Epidemiology of Severe Acute Respiratory Syndrome (SARS)" (World Health Organization, May 16, 2003), iris.who.int/bitstream/handle/10665/70863/WHO_CDS_CSR_GAR_2003.11_eng.pdf?sequence=1.

66. WHO Western Pacific Region, *SARS: How a Global Epidemic Was Stopped.*

67. Institute of Medicine Forum on Microbial Threats, *Learning from SARS: Preparing for the Next Disease Outbreak: Workshop Summary.*

68. Oluwaseun Sharomi and Tufail Malik, "Optimal Control in Epidemiology," *Annals of Operations Research* 227, no. 251 (2017): 55–71, researchgate.net/publication/276154126_Optimal_control_in_epidemiology.

69. Jie Hua et al., "A Visual Approach for the SARS (Severe Acute Respiratory Syndrome) Outbreak Data Analysis," *International Journal of Environmental Research and Public Health* 17, no. 11 (June 2020): 3973, doi.org/10.3390/ijerph17113973.

70. WHO Western Pacific Region, *SARS: How a Global Epidemic Was Stopped.*

71. WHO Western Pacific Region, *SARS: How a Global Epidemic Was Stopped.*

72. "Official Speeches and Statements About the 'People's War' on SARS," *Chinese Law & Government* 36, no. 6 (2003): 3–29, doi.org/10.2753/CLG0009-460936063.

73. World Health Organization, "Summary of Probable SARS Cases with Onset of Illness from 1 November 2002 to 31 July 2003"; Hua et al., "A Visual Approach for the SARS (Severe Acute Respiratory Syndrome) Outbreak Data Analysis."

74. Tomislav Svoboda et al., "Public Health Measures to Control the Spread of the Severe Acute Respiratory Syndrome during the Outbreak in Toronto," *New England Journal of Medicine* 350, no. 23 (June 3, 2004): 2352–61, doi.org/10.1056/NEJMoa032111.

75. World Health Organization, "Consensus Document on the Epidemiology of Severe Acute Respiratory Syndrome (SARS)."

76. World Health Organization, "Summary of Probable SARS Cases with Onset of Illness from 1 November 2002 to 31 July 2003."

77. Louise H. Taylor et al., "Risk Factors for Human Disease Emergence," *Philosophical Transactions of the Royal Society B: Biological Sciences* 356, no. 1411 (July 29, 2001): 983–89, doi.org/10.1098/rstb.2001.0888.

78. WHO Western Pacific Region, *SARS: How a Global Epidemic Was Stopped.*

79. WHO Western Pacific Region, *SARS: How a Global Epidemic Was Stopped.*

80. Michael Letko et al., "Bat-Borne Virus Diversity, Spillover and Emergence," *Nature Reviews Microbiology* 18, no. 8 (August 2020): 461–71, doi.org/10.1038/s41579-020-0394-z.

81. WHO Western Pacific Region, *SARS: How a Global Epidemic Was Stopped.*

CHAPTER 4: FOUNDATIONS

1. Bhushan Patwardhan et al, "Chapter 3—Concepts of Health and Disease," in *Integrative Approaches for Health*, ed. Bhushan Patwardhan, Gururaj

Mutalik, and Girish Tillu (Boston: Academic Press, 2015), 53–78, doi.org /10.1016/B978-0-12-801282-6.00003-6.

2. Centers for Disease Control and Prevention, "Social Determinants of Health," February 7, 2024, cdc.gov/health-disparities-hiv-std-tb-hepatitis /about/social-determinants-of-health.html.

3. Lisa F. Berkman et al, *Social Epidemiology* (New York: Oxford University Press, 2014); James P. Smith, "Unraveling the SES: Health Connection," *Population and Development Review* 30 (2004): 108–32, rand.org/content /dam/rand/pubs/reprints/2005/RAND_RP1170.pdf.

4. UN Inter-agency Group for Child Mortality Estimation, "Chad: Infant Mortality Rate" (UNICEF Data Warehouse, 2021), childmortality.org/all -cause-mortality/data?refArea=TCD; World Health Organization, "Life Expectancy at Birth (Years)" (The Global Health Observatory, 2019), who .int/data/gho/data/indicators/indicator-details/GHO/life-expectancy -at-birth-(years).

5. World Health Organization, "Life Expectancy at Birth (Years)."

6. UN Inter-agency Group for Child Mortality Estimation, "Algeria: Infant Mortality Rate—Total" (UNICEF Data Warehouse, 2021), childmortal ity.org/all-cause-mortality/data?indicator=MRY0&refArea=DZA.

7. World Health Organization, "Life Expectancy at Birth (Years)."

8. B. Tejada-Vera et al., "Life Expectancy at Birth for U.S. States and Census Tracts, 2010-2015," 2020, cdc.gov/nchs/data-visualization/life-expectancy /index.html.

9. County Health Rankings & Roadmaps, "Life Expectancy*," accessed February 2, 2024, countyhealthrankings.org/explore-health-rankings/county -health-rankings-model/health-outcomes/length-of-life/life-expectancy.

10. Julian D. De Lia et al., "Fetoscopic Laser Ablation of Placental Vessels in Severe Previable Twin-Twin Transfusion Syndrome," *American Journal of Obstetrics and Gynecology* 172, no. 4, Part 1 (April 1995): 1202–11, https:// doi.org/10.1016/0002-9378(95)91480-3; Julian E. De Lia et al., "Placental Surgery: A New Frontier," *Placenta* 14, no. 5 (September 1993): 477–85, https://doi.org/10.1016/S0143-4004(05)80201-2.

11. Latoya Hill et al., "Racial Disparities in Maternal and Infant Health: Current Status and Efforts to Address Them," *KFF* (blog), November 1, 2022, kff.org /racial-equity-and-health-policy/issue-brief/racial-disparities-in-maternal-and

-infant-health-current-status-and-efforts-to-address-them/; Carrie Wolfson et al., "Findings From Severe Maternal Morbidity Surveillance and Review in Maryland," *JAMA Network Open* 5, no. 11 (November 29, 2022), doi.org/10.1001/jamanetworkopen.2022.44077.

12. Eliseo J. Pérez-Stable and Monica Webb Hooper, "Acknowledgment of the Legacy of Racism and Discrimination," *Ethnicity & Disease* 31, no. Suppl 1 (2021): 289–92, ncbi.nlm.nih.gov/pmc/articles/PMC8143856/.

13. Carolyn B. Swope et al., "The Relationship of Historical Redlining with Present-Day Neighborhood Environmental and Health Outcomes: A Scoping Review and Conceptual Model," *Journal of Urban Health* 99, no. 6 (December 1, 2022): 959–83, doi.org/10.1007/s11524-022-00665-z.

14. Teshia G. Arambula Solomon et al., "The Generational Impact of Racism on Health: Voices from American Indian Communities," *Health Affairs* 41, no. 2 (February 2022), 281–88, healthaffairs.org/doi/10.1377/hlthaff.2021.01419.

15. Richard Gelting et al., "Water, Sanitation and Hygiene in Haiti: Past, Present, and Future," *The American Journal of Tropical Medicine and Hygiene* 89, no. 4 (October 9, 2013): 665–70, doi.org/10.4269/ajtmh.13-0217.

16. UNESCO, "Report by the Director-General on UNESCO's Cooperation with Haiti," Executive Board, 182nd (Paris, France: UNESCO, September 11, 2009), UNESCO Digital Library, https://unesdoc.unesco.org/ark:/48223/pf0000184036.

17. Inyoung Kang, "A List of Previous Disasters in Haiti, a Land All too Familiar with Hardship," *New York Times*, October 4, 2016, nytimes.com/2016/10/05/world/americas/haiti-hurricane-earthquake.html.

18. UN Country Team in Haiti, "Haiti: Food Crisis Report Jul 2008," ReliefWeb, July 22, 2008, reliefweb.int/report/haiti/haiti-food-crisis-report-jul-2008.

19. Adam Hochberg, "Air Traffic Over Haiti Is Crowded, Chaotic," NPR, January 15, 2010, sec. Latin America, npr.org/2010/01/15/122599423/air-traffic-over-haiti-is-crowded-chaotic.

20. Government of the Republic of Haiti, *Action Plan for National Recovery and Development of Haiti* (Haiti: Government of the Republic of Haiti, 2010), reliefweb.int/report/haiti/action-plan-national-recovery-and-development-haiti.

21. Richard Pallardy, "2010 Haiti Earthquake | Magnitude, Damage, Map, & Facts," in *Encyclopedia Britannica*, January 5, 2022, britannica.com/event /2010-Haiti-earthquake; Government of the Republic of Haiti, *Action Plan for National Recovery and Development of Haiti*.

22. Donatella Lippi and E. Gotuzzo, "The Greatest Steps towards the Discovery of *Vibrio Cholerae*," *Clinical Microbiology and Infection* 20, no. 3 (March 1, 2014): 191–95, doi.org/10.1111/1469-0691.12390; Gian Piero Carboni, "The Enigma of Pacini's *Vibrio Cholerae* Discovery," *Journal of Medical Microbiology* 70, no. 11 (November 2021), doi.org/10.1099/jmm.0.001450.

23. World Health Organization, "Cholera," December 11, 2023, who.int/news -room/fact-sheets/detail/cholera.

24. John D. Clemens et al., "Cholera," *Lancet* 390, no. 10101 (September 23, 2017): 1539–49, doi.org/10.1016/S0140-6736(17)30559-7.

25. Centers for Disease Control and Prevention, "Cholera: Causes and How It Spreads," April 12, 2022, cdc.gov/cholera/causes/.

26. G. F. Pyle, "The Diffusion of Cholera in the United States in the Nineteenth Century—Pyle—1969—Geographical Analysis—Wiley Online Library," *Geographical Analysis* 1, no. 1 (January 1969): 59–75, onlineli brary.wiley.com/doi/10.1111/j.1538-4632.1969.tb00605.x.

27. Centers for Disease Control and Prevention, "Five Basic Cholera Prevention Steps," July 25, 2023, cdc.gov/cholera/prevention/.

28. Dr. Robert D. Morris, *The Blue Death: The Intriguing Past and Present Danger of the Water You Drink*, reprint edition (Harper Perennial, 2008).

29. United Nations, "The Human Right to Water and Sanitation," accessed January 10, 2024, un.org/waterforlifedecade/human_right_to_water.shtml.

30. Mohammad Ali et al., "Updated Global Burden of Cholera in Endemic Countries," *PLoS Neglected Tropical Diseases* 9, no. 6 (June 4, 2015), doi .org/10.1371/journal.pntd.0003832.

31. United Nations, "The United Nations World Water Development Report 2023: Partnerships and Cooperation for Water" (Paris: UNESCO, 2023), unesdoc.unesco.org/ark:/48223/pf0000384655.

32. Abraham Ajayi and Stella I. Smith, "Recurrent Cholera Epidemics in Africa: Which Way Forward? A Literature Review," *Infection* 47, no. 3 (June 1, 2019): 341–49, doi.org/10.1007/s15010-018-1186-5; Angel N. Desai et al., "Infectious Disease Outbreaks among Forcibly Displaced Persons: An

Analysis of ProMED Reports 1996–2016," *Conflict and Health* 14, no. 1 (July 22, 2020): 49, doi.org/10.1186/s13031-020-00295-9.

33. WHO/UNICEF Joint Monitoring Programme (for Water Supply, Sanitation, and Hygiene), "State of the World's Drinking Water (October 2022)," accessed January 31, 2024, washdata.org/reports/state-worlds-drinking-water.

34. Renaud Piarroux et al., "Understanding the Cholera Epidemic, Haiti," *Emerging Infectious Diseases* 17, no. 7 (July 2011): 1161–68, doi.org/10.3201/eid1707.110059.

35. Alejandro Cravioto et al., "Final Report of the Independent Panel of Experts on the Cholera Outbreak in Haiti," Independent Panel Report (United Nations, May 6, 2011), reliefweb.int/report/haiti/final-report-independent-panel-experts-cholera-outbreak-haiti.

36. Jonathan M. Katz, "UN Probes Base as Source of Haiti Cholera Outbreak," Associated Press, October 28, 2010, medicalxpress.com/news/2010-10-probes-base-source-haiti-cholera.html.

37. Jonathan M. Katz, "Haiti Cholera Likely from UN Troops, Expert Says," Associated Press, December 7, 2010, nbcnews.com/id/wbna40553068; Jonathan M. Katz, "Experts Ask: Did U.N. Troops Infect Haiti?," Associated Press, November 3, 2010, nbcnews.com/health/health-news/experts-ask-did-u-n-troops-infect-haiti-flna1c9472818; Katz, "UN Probes Base as Source of Haiti Cholera Outbreak."

38. Katz, "UN Probes Base as Source of Haiti Cholera Outbreak."

39. Rene S. Hendriksen et al., "Population Genetics of *Vibrio Cholerae* from Nepal in 2010: Evidence on the Origin of the Haitian Outbreak," *mBio* 2, no. 4 (August 23, 2011): e00157-11, doi.org/10.1128/mBio.00157-11.

40. Cravioto et al., "Final Report of the Independent Panel of Experts on the Cholera Outbreak in Haiti."

41. Mariam Claeson and Ronald Waldman, "Cholera—Pandemic, Waterborne, 19th Century," in *Encyclopaedia Britannica*, accessed January 10, 2024, britannica.com/science/cholera/Cholera-through-history.

42. Direction d'Epidémiologie, des Laboratoires et de la Recherche (DELR), MSPP, "Rapport Du Réseau National de Surveillance Choléra: 3 Ème Semaine Épidémiologique 2020" (Haiti: Ministère Sante Publique et de la Population, 2020).

43. MSPP, "Rapport Du Réseau National de Surveillance Choléra."

44. MSPP, "Rapport Du Réseau National de Surveillance Choléra."

45. MSPP, "Rapport Du Réseau National de Surveillance Choléra."

46. Elizabeth C. Lee et al., "Achieving Coordinated National Immunity and Cholera Elimination in Haiti through Vaccination: A Modelling Study," *The Lancet Global Health* 8, no. 8 (August 1, 2020): e1081–89, doi.org /10.1016/S2214-109X(20)30310-7; MSPP, "Rapport Du Réseau National de Surveillance Choléra."

47. Camila Domonoske, "UN Admits Role in Haiti Cholera Outbreak That Has Killed Thousands," NPR, August 18, 2016, npr.org/sections/thetwo -way/2016/08/18/490468640/u-n-admits-role-in-haiti-cholera-outbreak -that-has-killed-thousands.

48. UN Human Rights Office of the High Commissioner, "UN Inaction Denies Justice for Haiti Cholera Victims, Say UN Experts," April 30, 2020, www.ohchr.org/en/press-releases/2020/04/un-inaction-denies-justice-haiti -cholera-victims-say-un-experts.

49. UN Haiti Cholera Response Multi-Partner Trust Fund, "UN Haiti Cholera Response Multi-Partner Trust Fund: 2020 Annual Report" (Office of the UN Secretary-General's Special Envoy for Haiti, 2021), 5, mptf.undp .org/factsheet/fund/CLH00.

50. UN Haiti Cholera Response Multi-Partner Trust Fund, "UN Haiti Cholera Response Multi-Partner Trust Fund: 2020 Annual Report," 22.

51. World Health Organization and UNICEF, "Haiti SDG6 Data," UN Water, n.d., sdg6data.org/en/country-or-area/Haiti.

52. World Health Organization and UNICEF, "Haiti SDG6 Data."

53. Pan American Health Organization/World Health Organization, "Cholera Epidemic in Haiti and the Dominican Republic," Situation Report 20, December 27, 2023), paho.org/en/file/138408/download?token=RAYjHEN8.

54. Sophia Campbell and David Wessel, "What Would It Cost to Replace All the Nation's Lead Water Pipes?," *Brookings Institution* (blog), May 13, 2021, brookings.edu/articles/what-would-it-cost-to-replace-all-the-nations-lead -water-pipes/.

55. US Environmental Protection Agency, "EPA's 7th Drinking Water Needs Survey and Assessment" (April 2023), epa.gov/ground-water-and-drinking -water/epas-7th-drinking-water-infrastructure-needs-survey-and-assessment.

56. Kathryn B. Egan et al., "Blood Lead Levels in U.S. Children Ages 1–11 Years, 1976–2016," *Environmental Health Perspectives* 129, no. 3 (n.d.): 037003, doi.org/10.1289/EHP7932; Marissa Hauptman et al., "Individual- and Community-Level Factors Associated with Detectable and Elevated Blood Lead Levels in US Children: Results From a National Clinical Laboratory," *JAMA Pediatrics* 175, no. 12 (December 1, 2021): 1252–60, doi.org/10.1001/jamapediatrics.2021.3518; Marissa Hauptman, Rebecca Bruccoleri, and Alan D. Woolf, "An Update on Childhood Lead Poisoning," *Clinical Pediatric Emergency Medicine* 18, no. 3 (September 2017): 181–92, doi.org/10.1016/j.cpem.2017.07.010.

57. Hauptman et al., "Individual- and Community-Level Factors Associated with Detectable and Elevated Blood Lead Levels in US Children."

58. US Environmental Protection Agency, Office of Land and Emergency Management, "Population Surrounding 1,881 Superfund Sites," Reports and Assessments, 2023, epa.gov/superfund/population-surrounding-1881 -superfund-sites.

59. US Environmental Protection Agency, "Aberdeen Proving Ground (Edgewood Area) Edgewood, MD Cleanup Activities," accessed December 3, 2023, https://cumulis.epa.gov/supercpad/SiteProfiles/index.cfm?fuseaction= second.cleanup&id=0300421.

60. Donald A. Henderson, "The Eradication of Smallpox—An Overview of the Past, Present, and Future," *Vaccine* 29, no. S4 (December 30, 2011): D7–9, doi.org/10.1016/j.vaccine.2011.06.080.

61. "Essential Programme on Immunization," accessed February 2, 2024, who .int/teams/immunization-vaccines-and-biologicals/essential-programme-on -immunization.

62. World Health Organization, "Supply Chain," n.d., who.int/teams/immu nization-vaccines-and-biologicals/essential-programme-on-immunization /supply-chain.

63. Vaccine-preventable Diseases and Immunization Programme et al., "Public Health Dispatch: Certification of Poliomyelitis Eradication—European Region, June 2002," *Morbidity and Mortality Weekly Report* 51, no. 26 (July 5, 2002): 572–74; Western Pacific Regional Office et al., "Public Health Dispatch: Certification of Poliomyelitis Eradication—Western Pacific Region, October 2000," *Morbidity and Mortality Weekly Report* 50, no. 1 (January 12, 2001): 1–3.

64. World Health Organization, "Essential Programme on Immunization."

65. Centers for Disease Control and Prevention, "2014–2016 Ebola Outbreak in West Africa," March 8, 2019, cdc.gov/ebola/outbreaks/index.html.

66. Alpha Forna et al., "Case Fatality Ratio Estimates for the 2013–2016 West African Ebola Epidemic: Application of Boosted Regression Trees for Imputation," *Clinical Infectious Diseases* 70, no. 12 (June 10, 2020): 2476–83, doi.org/10.1093/cid/ciz678.

67. Faisal Shuaib et al., "Ebola Virus Disease Outbreak—Nigeria, July–September 2014," *Morbidity and Mortality Weekly Report* 63, no. 39 (October 3, 2014): 867–72, pubmed.ncbi.nlm.nih.gov/25275332/.

68. Shuaib et al., "Ebola Virus Disease Outbreak—Nigeria, July–September 2014."

69. Elizabeth W. Etheridge, *Sentinel for Health: A History of the Centers for Disease Control and Prevention*, 1st ed. (Berkeley: University of California Press, 1992).

CHAPTER 5: TRUTH TELLING

1. Heather Cox Richardson, "About Letters from an American," 2024, heather coxrichardson.substack.com/about.

2. John M. Barry, *The Great Influenza: The Epic Story of the Deadliest Plague in History* (New York: Viking, 2004).

3. Stephen Corfidi, "A Brief History of the Storm Prediction Center," National Oceanic and Atmospheric Administration, n.d., spc.noaa.gov/history/early .html.

4. Corfidi, "A Brief History of the Storm Prediction Center."

5. Corfidi, "A Brief History of the Storm Prediction Center."

6. "IHR Emergency Committee on Novel Coronavirus (2019-nCoV)," accessed February 2, 2024, who.int/director-general/speeches/detail/who-director -general-s-statement-on-ihr-emergency-committee-on-novel-coronavirus -(2019-ncov).

7. Peter Felknor, *The Tri-State Tornado: The Story of America's Greatest Tornado Disaster* (iUniverse, 2004).

8. Ryan Presley, "1925 Tri-State Tornado—First-Hand Accounts," National Weather Service, n.d., weather.gov/pah/1925Tornado_fha.

9. Presley, "1925 Tri-State Tornado—First-Hand Accounts."

10. Presley, "1925 Tri-State Tornado—First-Hand Accounts."

11. Presley, "1925 Tri-State Tornado—First-Hand Accounts."

12. Felknor, *The Tri-State Tornado.*

13. "Wind-Driven Wall of Water Swept Gorham Where 90 Were Killed," *St. Louis Post-Dispatch,* March 20, 1925, quoted in Felknor, *The Tri-State Tornado.*

14. Presley, "1925 Tri-State Tornado—First-Hand Accounts."

15. Felknor, *The Tri-State Tornado.*

16. Marlene Bradford, "Historical Roots of Modern Tornado Forecasts and Warnings," *Weather and Forecasting* 14, no. 4 (August 1999): 484–91, doi .org/10.1175/1520-0434(1999)014<0484:HROMTF>2.0.CO;2.

17. Felknor, *The Tri-State Tornado.*

18. *Daily Illini,* "1,000 Die in Tornado: Southern Illinois Is Laid Waste as Fire Follows Storm," *Daily Illini* LIV, no. 156 (March 19, 1925): 1, genealogytrails .com/ill/perry/1925tornado.html.

19. *Daily Illini,* "1,000 Die in Tornado: Southern Illinois Is Laid Waste as Fire Follows Storm."

20. Felknor, *The Tri-State Tornado.*

21. NOAA's National Weather Service, "Tornado Threat Description," accessed December 30, 2023, weather.gov/mlb/tornado_threat.

22. Ryan Presley, "NOAA/NWS Tri-State Tornado Web Site—Startling Statistics," National Weather Service, n.d., weather.gov/pah/1925Tornado_ss.

23. NOAA's National Weather Service, "Tornado Threat Description."

24. Joseph G. Galway, "J. P. Finley: The First Severe Storms Forecaster," *Bulletin of the American Meteorological Society* 66, no. 11 (November 1985): 1389–95, journals.ametsoc.org/view/journals/bams/66/11/1520-0477_1985_066 _1389_jftfss_2_0_co_2.xml; Joseph G. Galway, "J. P. Finley: The First Severe Storms Forecaster," NOAA Technical Memorandum ERL NSSL-97 (Norman, OK: National Severe Storms Laboratory, November 1984), repos itory.library.noaa.gov/view/noaa/9484.

25. Rebecca Robbins Raines, "Signal Corps" (Washington, DC: Center of Military History, United States Army, 2005), history.army.mil/html/books /060/60-15-1/CMH_Pub_60-15-1.pdf.

26. Marlene Bradford, "Historical Roots of Modern Tornado Forecasts and Warnings," *Weather & Forecasting* 14, no. 4 (August 1999): 484, doi.org /10.1175/1520-0434(1999)014<0484:HROMTF>2.0.CO;2.

27. Galway, "J. P. Finley: The First Severe Storms Forecaster."

28. J. P. Finley, "Report of the Tornadoes of May 29 and 30, 1879, in Kansas, Nebraska, Missouri, and Iowa," Professional Papers of the Signal Service Prepared under the Direction of W. B. Hazen (Washington, DC: US Government Printing Office, 1881), 32.

29. Finley, "Report of the Tornadoes of May 29 and 30, 1879, in Kansas, Nebraska, Missouri, and Iowa," 28.

30. Galway, "J. P. Finley: The First Severe Storms Forecaster," 1390.

31. Galway, "J. P. Finley: The First Severe Storms Forecaster," 1390.

32. Galway, "J. P. Finley: The First Severe Storms Forecaster," 1390.

33. Galway, "J. P. Finley: The First Severe Storms Forecaster," 1390.

34. Bradford, "Historical Roots of Modern Tornado Forecasts and Warnings," August 1999; J. P. Finley, "Intelligence from American Scientific Stations," *Science* 3, no. 72 (1884): 766–68.

35. Bradford, "Historical Roots of Modern Tornado Forecasts and Warnings."

36. G. K. Gilbert, "Finley's Tornado Predictions," *American Meteorological Journal. A Monthly Review of Meteorology and Allied Branches of Study (1884–1896)* 1, no. 5 (September 1884): 166.

37. Bradford, "Historical Roots of Modern Tornado Forecasts and Warnings," 487.

38. Bradford, "Historical Roots of Modern Tornado Forecasts and Warnings," 489

39. Bradford, "Historical Roots of Modern Tornado Forecasts and Warnings," 489.

40. Bradford, "Historical Roots of Modern Tornado Forecasts and Warnings," 489.

41. Galway, Joseph. "J. P. Finley: The First Severe Storms Forecaster," NSSL Technical Memoranda. Norman, Oklahoma: National Oceanic and Atmospheric Administration, November 1984, 28. repository.library.noaa .gov/view/noaa/9484.

42. "Health and Human Services Briefing on the Coronavirus Outbreak," C-SPAN, February 25, 2020, c-span.org/video/?469708-1/health-human -services-briefing-coronavirus-outbreak.

43. Ebony Bowden, Carl Campanile, and Bruce Golding, "Worker at NYC Hospital Where Nurses Wear Trash Bags as Protection Dies from Coronavirus,"

New York Post, March 26, 2020, https://nypost.com/2020/03/25/worker
-at-nyc-hospital-where-nurses-wear-trash-bags-as-protection-dies-from
-coronavirus/.

44. D. F Johnson et al., "A Quantitative Assessment of the Efficacy of Surgical and N95 Masks to Filter Influenza Virus in Patients with Acute Influenza Infection," *Clinical Infectious Diseases* 49, no. 2 (July 15, 2009): 275–77, doi.org/10.1086/600041; Abhiteja Konda et al., "Aerosol Filtration Efficiency of Common Fabrics Used in Respiratory Cloth Masks," *ACS Nano* 14, no. 5 (May 26, 2020): 6339–47, doi.org/10.1021/acsnano.0c03252.

45. Ying Shan Doris Zhang et al., "'Responsible' or 'Strange?' Differences in Face Mask Attitudes and Use Between Chinese and Non-East Asian Canadians During COVID-19's First Wave," *Frontiers in Psychology* 13 (2022), frontiersin.org/articles/10.3389/fpsyg.2022.853830.

46. John P. Thornhill et al., "Monkeypox Virus Infection in Humans across 16 Countries—April–June 2022," *New England Journal of Medicine* 387, no. 8 (August 25, 2022): 679–91, doi.org/10.1056/NEJMoa2207323. ·

47. Bradford, "Historical Roots of Modern Tornado Forecasts and Warnings," 490.

48. Robert C. Miller and Charlie A. Crisp, "The First Operational Tornado Forecast Twenty Million to One," *Weather and Forecasting* 14, no. 4 (August 1, 1999): 479, doi.org/10.1175/1520-0434(1999)014<0479:TFOTFT> 2.0.CO;2.

49. Miller and Crisp, "The First Operational Tornado Forecast Twenty Million to One," 479.

50. James L. Crowder, "Tinker's Twin Twisters of 1948 and the Birth of Tornado Forecasting," *Chronicles of Oklahoma* 78, no. 3 (2000): 278–95, gateway.okh istory.org/ark:/67531/metadc2016810/.

51. Crowder, "Tinker's Twin Twisters of 1948 and the Birth of Tornado Forecasting," 282.

52. Crowder, "Tinker's Twin Twisters of 1948 and the Birth of Tornado Forecasting," 280.

53. Robert C. Miller, "The Unfriendly Sky," February 23, 1999, web.archive .org/web/20111101100002/http://www.nssl.noaa.gov/GoldenAnniversary /Historic.html.

54. Miller, "The Unfriendly Sky."

55. Robert C. Miller and Charlie A. Crisp, "Events Leading to Establishment of the United States Air Force Severe Weather Warning Center in February 1951," *Weather & Forecasting* 14, no. 4 (1998): 501.

56. Miller and Crisp, "The First Operational Tornado Forecast Twenty Million to One," 481.

57. Miller and Crisp, "The First Operational Tornado Forecast Twenty Million to One," 481.

58. Benjamin Haynes and Nancy E. Messonnier, "CDC Media Telebriefing: Update on COVID-19: For Immediate Release," press briefing audio recording (CDC, 2020), stacks.cdc.gov/view/cdc/85310/comingFrom Search.

59. Haynes and Messonnier, "CDC Media Telebriefing," loc. 2:45.

60. "Health and Human Services Briefing on the Coronavirus Outbreak."

61. Haynes and Messonnier, "CDC Media Telebriefing," loc. 7:54.

62. Gregg K. Vandekieft, "Breaking Bad News," *American Family Physician* 64, no. 12 (December 15, 2001): 1975–79, aafp.org/pubs/afp/issues/2001/1215/p1975.html.

63. Haynes and Messonnier, "CDC Media Telebriefing," loc. 8:22.

64. Tamar Klaiman, John D. Kraemer, and Michael A. Stoto, "Variability in School Closure Decisions in Response to 2009 H1N1: A Qualitative Systems Improvement Analysis," *BMC Public Health* 11 (February 1, 2011): 73, doi.org/10.1186/1471-2458-11-73.

65. National Weather Service, "Forecasts and Services," National Oceanic and Atmospheric Administration, accessed January 19, 2024, weather.gov/about/forecastsandservice.

66. Storm Prediction Center, "Apr 13, 2012 0600 UTC Day 2 Convective Outlook," accessed February 2, 2024, spc.noaa.gov/products/outlook/archive/2012/day2otlk_20120413_0600.html.

67. "Warnings Credited in Midwest Tornado Outbreak," *The Times Herald* (blog), April 16, 2012, timesherald.com/2012/04/16/warnings-credited-in-midwest-tornado-outbreak/.

68. National Oceanic and Atmospheric Administration Weather Program Office, "Social Science: Driving Research for Forecast Delivery and Decision-Making," National Oceanic and Atmospheric Administration, Weather Program Office, n.d., wpo.noaa.gov/social-science/.

69. National Oceanic and Atmospheric Administration Weather Program Office, "Weather Ready Quick Response Research," National Oceanic and Atmospheric Administration, Weather Program Office, n.d., wpo.noaa .gov/weather-ready-quick-response-research/.

70. National Oceanic and Atmospheric Administration Weather Program Office, "WPO System for Public Access to Research Knowledge (SPARK)," National Oceanic and Atmospheric Administration, accessed January 27, 2024, app.smartsheetgov.com/b/publish?EQBCT=42d85120190d459299 46c157a1b6b584.

CHAPTER 6: POLITICS

1. Courtney Bublé, "Dr. Fauci to Young Scientists: Follow the Science and 'Stay out of Politics,'" *Government Executive*, September 7, 2022, govexec .com/management/2022/09/dr-fauci-young-scientists-follow-science-stay -out-politics/376788/.

2. Olivia B. Waxman, "FBI Directors Are Appointed for 10-Year Terms. Here's Why," *Time*, May 10, 2017, time.com/4774610/james-comey-fbi-term-limit/.

3. The White House, "Memorandum on Restoring Trust in Government Through Scientific Integrity and Evidence-Based Policymaking," January 27, 2021, whitehouse.gov/briefing-room/presidential-actions/2021/01/27 /memorandum-on-restoring-trust-in-government-through-scientific -integrity-and-evidence-based-policymaking/.

4. David J. Sencer CDC Museum, "Past CDC Directors/Administrators," Centers for Disease Control and Prevention, June 26, 2023, cdc.gov/museum /history/pastdirectors.html.

5. Tom Griffin, "Calling the Shots: Dr. William Foege. The Man Who Helped Banish Smallpox from the Earth Is the 1994 Alumnus of the Year.," *University of Washington Magazine*, June 1994, washington.edu/alumni/columns /top10/calling_the_shots.html.

6. Jennifer Chapman and Justin K. Arnold, "Reye Syndrome," in *StatPearls* (Treasure Island, FL: StatPearls Publishing, 2024), ncbi.nlm.nih.gov /books/NBK526101/.

7. Ermias D. Belay et al., "Reye's Syndrome in the United States from 1981 through 1997," *New England Journal of Medicine* 340, no. 18 (May 6, 1999):

1377–82, doi.org/10.1056/NEJM199905063401801; Devra Lee Davis and Patricia Buffler, "Reduction of Deaths after Drug Labelling for Risk of Reye's Syndrome," *The Lancet,* originally published as Volume 2, Issue 8826, 340, no. 8826 (October 24, 1992): 1042, doi.org/10.1016/0140-6736(92) 93058-U.

8. Stephen B. Soumerai et al., "Effects of Professional and Media Warnings about the Association between Aspirin Use in Children and Reye's Syndrome," *The Milbank Quarterly* 70, no. 1 (1992): 155, doi.org/10.2307 /3350088.

9. Ronald J. Waldman et al., "Aspirin as a Risk Factor in Reye's Syndrome," *The Journal of the American Medical Association* 247, no. 22 (June 11, 1982): 3089–94, doi.org/10.1001/jama.1982.03320470035029; Belay et al., "Reye's Syndrome in the United States from 1981 through 1997."

10. Davis and Buffler, "Reduction of Deaths after Drug Labelling for Risk of Reye's Syndrome"; Eugene S. Hurwitz et al., "National Surveillance for Reye Syndrome: A Five-Year Review," *Pediatrics* 70, no. 6 (December 1, 1982): 895–900, doi.org/10.1542/peds.70.6.895.

11. Griffin, "Calling the Shots."

12. Soumerai et al., "Effects of Professional and Media Warnings about the Association between Aspirin Use in Children and Reye's Syndrome."

13. Soumerai et al. "Effects of Professional and Media Warnings about the Association between Aspirin Use in Children and Reye's Syndrome"; Irvin Molotsky, "Critics Say F.D.A. Is Unsafe in Reagan Era," *New York Times,* January 4, 1987, nytimes.com/1987/01/04/weekinreview/critics-say-fda-is -unsafe-in-reagan-era.html.

14. Waldman et al., "Aspirin as a Risk Factor in Reye's Syndrome."

15. Associated Press, "Aspirin Labels to Warn About Reye Syndrome," *New York Times,* March 8, 1986, National edition, sec. US.

16. Belay et al., "Reye's Syndrome in the United States from 1981 through 1997"; Cristine Russell, "Nader Health Unit Criticizes FDA in Delay on Aspirin," *Washington Post,* April 28, 1982, washingtonpost.com/archive /politics/1982/04/28/nader-health-unit-criticizes-fda-in-delay-on-aspirin /fabfa4e2-e869-4fce-b605-72052b1e084d/.

17. Davis and Buffler, "Reduction of Deaths after Drug Labelling for Risk of Reye's Syndrome."

18. Griffin, "Calling the Shots."

19. Kavya Sekar, "National Institutes of Health (NIH) Funding FY1996 -FY2023" (Washington, DC: Congressional Research Service, March 8, 2023), crsreports.congress.gov/product/pdf/R/R43341/45.

20. Kavya Sekar, "Centers for Disease Control and Prevention (CDC) Funding Overview" (Washington, DC: Congressional Research Service, March 28, 2023), crsreports.congress.gov/product/pdf/R/R47207.

21. Sekar, "National Institutes of Health (NIH) Funding FY1996-FY2023."

22. National Institute of Diabetes and Digestive and Kidney Diseases, "Celiac Disease," National Institute of Diabetes and Digestive and Kidney Diseases, accessed February 2, 2024, niddk.nih.gov/health-information/digestive -diseases/celiac-disease.

23. National Institutes of Health, "Finding a Clinical Trial," National Institutes of Health (NIH), November 6, 2018, nih.gov/health-information/nih -clinical-research-trials-you/finding-clinical-trial; National Library of Medicine, "ClinicalTrials.Gov," National Library of Medicine, n.d., clinical trials.gov/.

24. Sekar, "Centers for Disease Control and Prevention (CDC) Funding Overview."

25. This quote has been lightly edited for clarity. The full quote is: "So, if you want to play you got to be in the game and the game is not played in Washington, at least—I mean, the game is not played in Atlanta unless, you know, you're a fan of the baseball team there." Dr. Gerberding made the comment in a 2023 interview with the Center for Strategic and International Studies; Stephen J. Morrison and Katherine E. Bliss, "Report Launch: Building the CDC the Country Needs," January 17, 2023, csis.org/analy sis/report-launch-building-cdc-country-needs.

26. Morrison and Bliss, "Report Launch: Building the CDC the Country Needs."

27. Paul M. Sharp and Beatrice H. Hahn, "Origins of HIV and the AIDS Pandemic," *Cold Spring Harbor Perspectives in Medicine:* 1, no. 1 (September 2011), doi.org/10.1101/cshperspect.a006841.

28. Institute for Health Metrics and Evaluation, Global Burden of Disease (2019)—processed by Our World in Data, "Share of All Deaths Caused by HIV/AIDS, 1990 to 2019," 2019, ourworldindata.org/hiv-aids.

29. D. J. Hunter, "AIDS in Sub-Saharan Africa: The Epidemiology of Heterosexual Transmission and the Prospects for Prevention," *Epidemiology* 4, no. 1 (January 1993): 63–72, jstor.org/stable/3702986.

30. Alice Tseng, Jason Seet, and Elizabeth J. Phillips, "The Evolution of Three Decades of Antiretroviral Therapy: Challenges, Triumphs and the Promise of the Future," *British Journal of Clinical Pharmacology* 79, no. 2 (February 2015): 182–94, doi.org/10.1111/bcp.12403.

31. Myron S. Cohen et al., "Antiretroviral Therapy for the Prevention of HIV-1 Transmission," *The New England Journal of Medicine* 375, no. 9 (September 1, 2016): 830–39, doi.org/10.1056/NEJMoa1600693.

32. "Africa: Life Expectancy, 07/03/00," accessed January 30, 2024, africa.upenn.edu/Urgent_Action/apic-070300.html.

33. World Health Organization, "Global AIDS Epidemic Shows No Sign of Abating; Highest Number of HIV Infections and Deaths Ever," November 25, 2003, who.int/japan/news/detail-global/25-11-2003-global-aids-epidemic-shows-no-sign-of-abating-highest-number-of-hiv-infections-and-deaths-ever; Crystal Cazier and Andrew Kaufman, "An Oral History of PEPFAR: How a 'Dream Big' Partnership Is Saving the Lives of Millions," George W. Bush Presidential Center, February 24, 2023, bushcenter.org/publications/an-oral-history-of-pepfar-how-a-dream-big-partnership-is-saving-the-lives-of-millions/.

34. Centers for Disease Control and Prevention, "Introduction of Routine HIV Testing in Prenatal Care—Botswana, 2004," *Morbidity and Mortality Weekly Report* 53, no. 46 (November 26, 2004): 1083–86, jamanetwork.com/journals/jama/fullarticle/200157.

35. Cazier and Kaufman, "An Oral History of PEPFAR."

36. George W. Bush, *Decision Points* (New York: Crown, 2010).

37. "Leading the Fight Against Global HIV/AIDS" (Congressional Record 149, no. 72, May 14, 2003), congress.gov/congressional-record/volume-149/issue-72/senate-section/article/S6249-1.

38. Cazier and Kaufman, "An Oral History of PEPFAR."

39. Todd Summers and Jennifer Kates, "Trends in U.S. Government Funding for HIV/AIDS Fiscal Years 1981 to 2004" (San Francisco, CA: The Henry J. Kaiser Family Foundation, 2004), kff.org/global-health-policy/issue-brief/issue-brief-trends-in-u-s-government/.

40. Cazier and Kaufman, "An Oral History of PEPFAR."

41. Condoleezza Rice, *No Higher Honor: A Memoir of My Years in Washington* (New York: Crown, 2011).

42. Cazier and Kaufman, "An Oral History of PEPFAR."

43. Office of the Press Secretary, "Fact Sheet: The President's Emergency Plan for AIDS Relief," The White House, accessed February 2, 2024, georgew bush-whitehouse.archives.gov/news/releases/2003/01/20030129-1 .html.

44. Office of the Press Secretary, "President Bush Discusses World AIDS Day," The White House, accessed February 2, 2024, georgewbush-whitehouse .archives.gov/news/releases/2007/11/20071130-4.html.

45. Office of the Press Secretary, "Fact Sheet: President Bush Announces Five-Year, $30 Billion HIV/AIDS Plan," The White House, May 30, 2007, georgewbush-whitehouse.archives.gov/news/releases/2007/05 /20070530-5.html.

46. Cazier and Kaufman, "An Oral History of PEPFAR."

47. PEPFAR, "PEPFAR 2022 Annual Report to Congress" (Washington, DC: PEPFAR, 2022), state.gov/wp-content/uploads/2022/05/PEPFAR 2022.pdf.

48. J. S. Santelli et al., "Abstinence Promotion Under PEPFAR: The Shifting Focus of HIV Prevention for Youth," *Global Public Health* 8, no. 1 (2013): 1–12, tandfonline.com/doi/full/10.1080/17441692.2012.759609.

49. Human Rights Watch, "Access to Condoms and HIV/AIDS Information: A Global Health and Human Rights Concern," December 2004, hrw.org /legacy/backgrounder/hivaids/condoms1204/condoms1204.pdf.

50. Jessica L. Prodger et al., "How Does Voluntary Medical Male Circumcision Reduce HIV Risk?," *Current HIV/AIDS Reports* 19, no. 6 (December 2022): 484–90, doi.org/10.1007/s11904-022-00634-w.

51. The White House, "Statement from President Joe Biden on the 20th Anniversary of the U.S. President's Emergency Plan for AIDS Relief (PEPFAR)," January 28, 2023, whitehouse.gov/briefing-room/statements-releases /2023/01/28/statement-from-president-joe-biden-on-the-20th-anniversary -of-the-u-s-presidents-emergency-plan-for-aids-relief-pepfar/.

52. PEPFAR, "Latest Global Program Results," (Washington, DC: PEPFAR, December 2022), state.gov/wp-content/uploads/2022/11/PEPFAR-Latest -Global-Results_December-2022.pdf.

CHAPTER 7: COMMITMENTS

1. NASA (National Aeronautics and Space Administration), "Planetary Protection," December 15, 2023, sma.nasa.gov/sma-disciplines/planetary -protection.

2. Gayane A. Kazarians et al., "The Evolution of Planetary Protection Implementation on Mars Landed Missions," in 2017 IEEE Aerospace Conference, 1–20, 2017.

3. NASA, "Double Asteroid Redirection Test (DART)," science.nasa.gov /mission/dart/.

4. NASA, "DART's Impact with Asteroid Dimorphos" (Official NASA Broadcast), September 26, 2022. youtube.com/watch?v=4RA8Tfa6Sck.

5. NASA, "NASA HQ Planetary Protection Officer," August 14, 2017, astro biology.nasa.gov/careers-employment/nasa-hq-planetary-protection -officer/.

6. Erin Biba, "Meet NASA's One and Only Planetary Protection Officer," *Scientific American*, October 1, 2014, scientificamerican.com/article/meet -nasa-s-one-and-only-planetary-protection-officer/.

7. Alexander Koch et al., "Earth System Impacts of the European Arrival and Great Dying in the Americas after 1492," *Quaternary Science Reviews* 207 (March 1, 2019): 13–36, doi.org/10.1016/j.quascirev.2018.12.004.

8. US Census Bureau, "U.S. Population Estimated at 335,893,238 on Jan. 1, 2024," Census.gov, December 28, 2023, census.gov/library/stories/2023 /12/happy-new-year-2024.html.

9. World Population Review, "US States—Ranked by Population 2024," 2024, worldpopulationreview.com/states.

10. Survival International, "Uncontacted Tribes and the Right to Self-Determination," https://www.ohchr.org/sites/default/files/documents/issues /indigenouspeoples/cfi/submmissionselfdetermination/subm-self-determi nation-under-indi-peop-ngos-survival-international.pdf.

11. Kyle Harper, *Plagues Upon the Earth* (Princeton, New Jersey: Princeton University Press, 2021).

12. Harper, *Plagues Upon the Earth.*

13. Clifford R. Backman, *The Worlds of Medieval Europe*, Fourth Edition (New York: Oxford University Press, 2021).

14. Samuel Cohn, "The Black Death and Consequences for Labor," *Labor* 2023, https://read.dukeupress.edu/labor/article-pdf/20/2/14/1859354/14cohn.pdf.

15. World Health Organization Regional Office for Africa, "Plague Outbreak Madagascar External Situation Report 12," December 2017, afro.who.int /health-topics/plague/plague-outbreak-situation-reports.

16. Centers for Disease Control and Prevention, "Maps and Statistics: Plague in the United States," November 16, 2022, cdc.gov/plague/maps -statistics/.

17. NASA, "Planetary Protection," Office of Safety and Mission Assurance, NASA, December 15, 2023, sma.nasa.gov/sma-disciplines/planetary -protection.

18. John D. Rummel, "Planetary Exploration in the Time of Astrobiology: Protecting against Biological Contamination," *Proceedings of the National Academy of Sciences* 98, no. 5 (February 27, 2001): 2128–31, doi.org/10.1073/pnas .061021398.

19. Soviet and American scientists collaborated in other domains, too. Both countries were instrumental to the decision to eradicate smallpox and were key leaders in its successful completion.

20. "Joshua Lederberg: Biographical Overview," Profiles in Science, National Library of Medicine, March 12, 2019, profiles.nlm.nih.gov/spotlight/bb /feature/biographical-overview.

21. Lederberg was a leader in US health security policy for decades, including on matters of biosecurity. In August of 2001, shortly before the September 11th attacks and the anthrax attacks on the United States, Lederberg warned the Senate Committee on Foreign Relations of the threat of biological threats and terrorism, calling it, "probably the most perplexing and gravest security challenge we face." This insight proved prescient—particularly his warning that, "[responding to bioterrorism] entails coordination of local, state, and federal assets and jurisdictions and the intersection of law enforcement, national security, and public health. A time of crisis is not ideal for debates over responsibility, authority, and funding." When the anthrax attacks began one month later, the difficulties that jurisdictions had in coordinating the criminal investigation, public health investigation, and outbreak response were a central challenge. National Academy of Sciences, "Testimony of Joshua Lederberg, Ph.D.," in *Biological Threats and Terrorism: Assessing the Science and Response Capabilities: Workshop Summary* (Washington, DC: National Academies Press, 2002), 236–37, ncbi.nlm.nih.gov /books/NBK98409/.

22. Michael Meltzer, *When Biospheres Collide: A History of NASA's Planetary Protection Programs* (Washington, DC: US Government Printing Office, 2012), nasa.gov/history/history-publications-and-resources/nasa-history-series /when-biospheres-collide/.

23. Meltzer, *When Biospheres Collide.*

24. Meltzer, *When Biospheres Collide.*

25. Norman H. Horowitz, "Letter from Norman H. Horowitz to Joshua Lederberg," January 30, 1960, Box: 11. Folder: 54, The Joshua Lederberg Papers, National Library of Medicine, profiles.nlm.nih.gov/spotlight/bb/catalog/nlm: nlmuid-101584906X6396-doc.

26. Meltzer, *When Biospheres Collide.*

27. Meltzer, *When Biospheres Collide*, chaps. 2 & 3.

28. John Uri, "60 Years Ago: Luna 2 Makes Impact in Moon Race," *NASA History* (blog), September 12, 2019, nasa.gov/history/60-years-ago-luna-2-makes -impact-in-moon-race/.

29. Meltzer, *When Biospheres Collide*, 62.

30. Meltzer, *When Biospheres Collide.*

31. NASA, "Apollo 11: The Moon Landing," 11, accessed February 2, 2024, https://airandspace.si.edu/explore/stories/apollo-11-moon-landing.

32. Meltzer, *When Biospheres Collide.*

33. SpaceFund, "Launch Database," October 12, 2022, spacefund.com/launch -database/.

34. Personal correspondence with a NASA official. The interview was conducted in October 2021 on a not-for-attribution basis.

35. United Nations Office for Outer Space Affairs, "Treaty on Principles Governing the Activities of States in the Exploration and Use of Outer Space, Including the Moon and Other Celestial Bodies," accessed January 23, 2024, unoosa.org/oosa/en/ourwork/spacelaw/treaties/introouterspacetreaty.html.

CHAPTER 8: SURPRISES

1. Donald Rumsfeld, *Known and Unknown: A Memoir* (New York: Sentinel, 2011).

2. Center for Arms Control and Non-Proliferation, "Federal Funding for Biological Weapons Prevention and Defense, Fiscal Years 2001 to 2008," June 6, 2007, armscontrolcenter.org/federal-funding-for-biological-weapons

-prevention-and-defense-fiscal-years-2001-to-2008/; Mahendra Pal et al., "An Overview on Biological Weapons and Bioterrorism," *American Journal of Biomedical Research* 5, no. 2 (2017): 24–34, pubs.sciepub.com/ajbr/5/2/2/index.html.

3. United Nations Office for Disarmament Affairs, "Biological Weapons Convention," n.d., disarmament.unoda.org/biological-weapons/.

4. Daniel Feakes, "The Biological Weapons Convention," *Revue Scientifique et Technique (International Office of Epizootics)* 36, no. 2 (August 2017): 621–28, doi.org/10.20506/rst.36.2.2679.

5. C. J. Davis, "Nuclear Blindness: An Overview of the Biological Weapons Programs of the Former Soviet Union and Iraq," *Emerging Infectious Diseases* 5, no. 4 (1999): 509–12, ncbi.nlm.nih.gov/pmc/articles/PMC2627746/.

6. Elizabeth Philipp et al., "North Korea's Biological Weapons Program: The Known and Unknown" (Boston, MA: Belfer Center for Science and International Affairs, Harvard Kennedy School, 2017), belfercenter.org/publication/north-koreas-biological-weapons-program-known-and-unknown.

7. Hugh B. Urban, *Zorba the Buddha: Sex, Spirituality, and Capitalism in the Global Osho Movement* (Berkeley: University of California Press, 2016).

8. Associated Press, "Followers of Guru May Start New City," *Eugene Register-Guard*, October 9, 1981, sec. D, Google News Archive, news.google.com/newspapers?id=dbRQAAAAIBAJ&sjid=VOIDAAAAIBAJ&pg=6551%2C2443993.

9. Hugh B. Urban, "Zorba the Buddha: Capitalism, Charisma and the Cult of Bhagwan Shree Rajneesh," *Religion* 26, no. 2 (April 1996): 161–82, https://doi.org/10.1006/reli.1996.0013.

10. Population Research Center, "Population Estimates for Oregon 1980–1989," Oregon Population Estimates and Reports (Portland, OR: Portland State University, 1990), PDXScholar, archives.pdx.edu/ds/psu/35818.

11. W. Seth Carus, "Bioterrorism and Biocrimes: The Illicit Use of Biological Agents Since 1900," working paper (Washington, DC: Center for Counterproliferation Research, National Defense University, February 2001), irp.fas.org/threat/cbw/carus.pdf.

12. "Oregon Voting Laws Part of the Problem," *Longview Daily News*, September 27, 1984.

13. Scott Keyes, "A Strange but True Tale of Voter Fraud and Bioterrorism,"

Atlantic, June 10, 2014, theatlantic.com/politics/archive/2014/06/a-strange
-but-true-tale-of-voter-fraud-and-bioterrorism/372445/; Carus, "Bioterrorism
and Biocrimes."

14. Keyes, "A Strange but True Tale of Voter Fraud and Bioterrorism."

15. Carus, "Bioterrorism and Biocrimes."

16. Carus, "Bioterrorism and Biocrimes," 31.

17. Carus, "Bioterrorism and Biocrimes."

18. Associated Press, "Bhagwan Shree Rajneesh, Commune Leader, Dies,"
Washington Post, January 20, 1990, washingtonpost.com/archive/local/1990
/01/20/bhagwan-shree-rajneesh-commune-leader-dies/fe872902-bfb0
-49ec-b875-17e4ab23b9df/.

19. Jarold Ramsey, "City of Antelope and Muddy Ranch," Oregon Encyclopedia, February 25, 2022, oregonencyclopedia.org/articles/city_of_antelope
_muddy_ranch/.

20. Donald A. Henderson, prepared statement, "Bioterrorism—Domestic Weapons of Mass Destruction: Joint Hearing before the Subcommittee on Labor,
Health, and Human Services, and Education, and Related Agencies: Committee on Appropriations and the Committee on Veterans' Affairs, United
States Senate, 106th Congress, First Session, Special Hearing," Senate Hearing 106-352, U.S. Government Publishing Office, congress.gov/event
/106th-congress/senate-event/LC19715/text.

21. Henderson, prepared statement, "Bioterrorism—Domestic Weapons of Mass
Destruction."

22. William R. Clark, "The Ultimate Bioterrorist: Mother Nature!," in *Bracing
for Armageddon?: The Science and Politics of Bioterrorism in America* (New
York: Oxford Academic, 2008), doi.org/10.1093/acprof:oso/9780195336214
.003.0005.

23. Johns Hopkins Bloomberg School of Public Health, "Donald Ainslie Henderson MD, MPH '60: In Memoriam," Johns Hopkins Bloomberg School
of Public Health, 2016, publichealth.jhu.edu/about/history/in-memoriam
/donald-a-henderson.

24. Scott Shane, "Portrait Emerges of Anthrax Suspect's Troubled Life," *New
York Times*, January 3, 2009, nytimes.com/2009/01/04/us/04anthrax.html.

25. Centers for Disease Control and Prevention, "What Is Anthrax?," February 15, 2022, cdc.gov/anthrax/about/index.html.

26. Joseph P. Wood et al., "Environmental Persistence of *Bacillus anthracis* and *Bacillus subtilis* Spores," *PLOS ONE* 10, no. 9 (September 15, 2015): e0138083, doi.org/10.1371/journal.pone.0138083.

27. Morton N. Swartz, "Recognition and Management of Anthrax—An Update," *New England Journal of Medicine* 345, no. 22 (November 29, 2001): 1621–26, doi.org/10.1056/NEJMra012892.

28. Centers for Disease Control and Prevention, "Types of Anthrax," November 19, 2020, cdc.gov/anthrax/about/.

29. Centers for Disease Control and Prevention, "Gastrointestinal Anthrax after an Animal-Hide Drumming Event—New Hampshire and Massachusetts, 2009," *Morbidity and Mortality Weekly Report* 59, no. 28 (July 23, 2010): 872–77, pubmed.ncbi.nlm.nih.gov/20651643/.

30. Swartz, "Recognition and Management of Anthrax—An Update."

31. Centers for Disease Control and Prevention, "Types of Anthrax."

32. Ellen A. Spotts Whitney et al., "Inactivation of *Bacillus anthracis* Spores," *Emerging Infectious Diseases* 9, no. 6 (June 2003): 623–27, doi.org/10.3201/eid0906.020377.

33. Robert A. Wampler and Thomas S. Blanton, eds., "Volume 5: Anthrax at Sverdlovsk, 1979" (Washington, DC: National Security Archive, George Washington University, November 15, 2001), nsarchive2.gwu.edu/NSAEBB/NSAEBB61/#doc1.

34. Raymond A. Zilinskas, "The Soviet Biological Weapons Program and Its Legacy in Today's Russia" (Washington, DC: National Defense University Press, July 2016), https://bwc1972.org/wp-content/uploads/2016/10/The-Soviet-Biological-Weapons-Program-and-Its-Legacy-in-Todays-Russia.pdf; Matthew Meselson et al., "The Sverdlovsk Anthrax Outbreak of 1979," *Science* 266, no. 5188 (1994): 1202–8.

35. Ken Alibek and Stephen Handelman, *Biohazard: The Chilling True Story of the Largest Covert Biological Weapons Program in the World—Told from the Inside by the Man Who Ran It* (New York: Delta, 2014).

36. Meselson et al., "The Sverdlovsk Anthrax Outbreak of 1979."

37. Ari Fleischer, "Press Briefing by Ari Fleischer," White House Press Briefing, Washington, DC, October 4, 2001, georgewbush-whitehouse.archives.gov/news/releases/2001/10/20011004-12.html.

38. Eric Lipton and Kirk Johnson, "A Nation Challenged: The Anthrax Trail;

Tracking Bioterror's Tangled Course," *New York Times*, December 26, 2001, nytimes.com/2001/12/26/us/a-nation-challenged-the-anthrax-trail-tracking -bioterror-s-tangled-course.html.

39. Fleischer, "Press Briefing by Ari Fleischer."

40. Centers for Disease Control and Prevention, "CERC: Crisis Communication Plans," 2014, 31, emergency.cdc.gov/cerc/ppt/CERC_Crisis_Communication _Plans.pdf.

41. Lawrence K. Altman, "The Doctor's World; C.D.C. Team Tackles Anthrax," *New York Times*, October 16, 2001, nytimes.com/2001/10/16/science /the-doctor-s-world-cdc-team-tackles-anthrax.html.

42. David Barstow, "Anthrax Found in NBC News Aide," *New York Times*, October 12, 2001, nytimes.com/2001/10/12/national/anthrax-found-in-nbc -news-aide.html.

43. Barstow, "Anthrax Found in NBC News Aide"; Altman, "The Doctor's World."

44. US Department of Justice, "Amerithrax Investigative Summary" (Washington, DC: US Department of Justice, February 19, 2010), justice.gov /archive/amerithrax/docs/amx-investigative-summary.pdf.

45. Centers for Disease Control and Prevention, "Update: Investigation of Bioterrorism-Related Anthrax and Adverse Events from Antimicrobial Prophylaxis," *Morbidity and Mortality Weekly Report* 50, no. 44 (November 9, 2001): 973–76; Government Accountability Office, "Bioterrorism: Public Health Response to Anthrax Incidents of 2001" (Washington, DC: US Government Accountability Office, October 2003), gao.gov/assets/gao -04-152.pdf.

46. US Department of Justice, "Amerithrax Investigative Summary."

47. US Department of Justice, "Amerithrax Investigative Summary."

48. Lipton and Johnson, "A Nation Challenged."

49. US Department of Justice, "Amerithrax Investigative Summary."

50. Smithsonian National Postal Museum, "Anthrax," n.d., postalmuseum .si.edu/exhibition/behind-the-badge-case-histories-dangerous-mail/an thrax.

51. US Department of Justice, "Amerithrax Investigative Summary."

52. Kelli Arena, "Hatfill Ticketed in Altercation with FBI Agent," CNN Washington Bureau, May 19, 2003, cnn.com/2003/US/05/19/hatfill/.

53. Nicholas D. Kristof, "Anthrax? The F.B.I. Yawns," *New York Times*, July 2, 2002, sec. Opinion, nytimes.com/2002/07/02/opinion/anthrax-the-fbi-yawns.html.

54. ABC News, "DOJ Settles Hatfill Suit for $5.8 Million," ABC News, accessed January 26, 2024, abcnews.go.com/TheLaw/DOJ/story?id=5264759&page=1.

55. Shane, "Portrait Emerges of Anthrax Suspect's Troubled Life."

56. Department of Justice, "Transcript of Amerithrax Investigation Press Conference," August 6, 2008, justice.gov/archive/opa/pr/2008/August/08-opa-697.html.

57. W. Seth Carus, "RISE, the Rajneeshees, Aum Shinrikyo and Bruce Ivins," in *Biological Threats in the 21st Century*, ed. Filippa Lentzos (London: Imperial College Press, 2016), 171–97.

58. Government Accountability Office, "Capitol Hill Anthrax Incident: EPA's Cleanup Was Successful; Opportunities Exist to Enhance Contract Oversight," Report to the Chairman, Committee on Finance, US Senate (Washington, DC: US Government Accountability Office, June 2003), gao.gov/assets/gao-03-686.pdf.

59. Donald A. Henderson, "3. Public Health Preparedness," American Association for the Advancement of Science, n.d., aaas.org/sites/default/files/stvwch3.pdf.

60. Monika Evstatieva, "A Revamped Strategic National Stockpile Still Can't Match the Pandemic's Latest Surge," NPR, November 23, 2020, sec. Investigations, npr.org/2020/11/23/937978556/a-revamped-strategic-national-stockpile-still-cant-match-the-pandemics-latest-su.

61. Government Accountability Office, "Biodefense: DHS Exploring New Methods to Replace BioWatch and Could Benefit from Additional Guidance" (Washington, DC: US Government Accountability Office, May 20, 2021), gao.gov/products/gao-21-292.

62. Government Accountability Office, "Biosurveillance: DHS Should Not Pursue BioWatch Upgrades or Enhancements Until System Capabilities Are Established" (Washington, DC: U.S. Government Accountability Office, October 23, 2015), gao.gov/products/gao-16-99.

63. Carus, "Bioterrorism and Biocrimes."

64. Erwin Schrödinger, "The Present Situation in Quantum Mechanics: A

Translation of Schrödinger's 'Cat Paradox' Paper," trans. John D. Trimmer, *Proceedings of the American Philosophical Society* 124 (1980): 323–38.

65. Nahida Chakhtoura et al., "Zika Virus: A Public Health Perspective," *Current Opinion in Obstetrics & Gynecology* 30, no. 2 (April 2018): 116–22, doi.org/10.1097/GCO.0000000000000440.

CHAPTER 9: TECHNOLOGIES

1. Candice L. Odgers et al., "Screen Time, Social Media Use, and Adolescent Development," *Annual Review of Developmental Psychology* 2, no. 1 (2020): 485–502, doi.org/10.1146/annurev-devpsych-121318-084815.

2. Tobias Dienlin and Niklas Johannes, "The Impact of Digital Technology Use on Adolescent Well-Being," *Dialogues in Clinical Neuroscience* 22, no. 2 (June 30, 2020): 135–42, doi.org/10.31887/DCNS.2020.22.2/tdienlin; Yi-Ju Wu et al., "A Systematic Review of Recent Research on Adolescent Social Connectedness and Mental Health with Internet Technology Use," *Adolescent Research Review* 1, no. 2 (June 1, 2016): 153–62, doi.org/10.1007/s40894-015-0013-9.

3. Karen M. Johnson-Weiner, "Technological Diversity and Cultural Change among Contemporary Amish Groups," *Mennonite Quarterly Review* 88, no. 1 (January 1, 2014): 5–23, goshen.edu/wp-content/uploads/sites/75/2016/06/Jan14Johnson.pdf.

4. Florida Keys Mosquito District, "Town Hall Meeting on Genetically Modified Mosquitoes," Key West, Florida, 2012, youtube.com/watch?v=5WQkM-yc7QI.

5. Florida Keys Mosquito Control District, "Keys Mosquitoes," 2021, keysmosquito.org/keys-mosquitoes/; Anna Maria Barry-Jester, "Small Island, Big Experiment," *FiveThirtyEight* (blog), October 18, 2016, fivethirtyeight.com/features/zika-mosquito-florida-vote/.

6. James J. Sejvar, "West Nile Virus: An Historical Overview," *The Ochsner Journal* 5, no. 3 (2003): 6–10, ncbi.nlm.nih.gov/pmc/articles/PMC3111838/.

7. André B. B. Wilke et al., "Ornamental Bromeliads of Miami-Dade County, Florida Are Important Breeding Sites for *Aedes aegypti* (Diptera: Culicidae)," *Parasites & Vectors* 11, no. 1 (May 17, 2018): 283, doi.org/10.1186/s13071-018-2866-9.

8. Centers for Disease Control and Prevention, "Life Cycle of *Aedes aegypti* and *Ae. albopictus* Mosquitoes," June 21, 2022, cdc.gov/mosquitoes/about /life-cycle-of-aedes-mosquitoes.html.

9. Michelle Bialeck, "Mosquito Control in the Florida Keys," *Scientific American Blog Network* (blog), April 11, 2012, scientificamerican.com/blog /guest-blog/mosquito-control-in-the-florida-keys/.

10. Barry-Jester, "Small Island, Big Experiment."

11. Casey Parker et al., "Baseline Susceptibility Status of Florida Populations of *Aedes aegypti* (Diptera: Culicidae) and Aedes albopictus," *Journal of Medical Entomology* 57, no. 5 (September 7, 2020): 1550–59, doi.org/10 .1093/jme/tjaa068.

12. Oxitec, "Our Technology," accessed February 2, 2024, oxitec.com/en/our -technology.

13. Danilo O. Carvalho et al., "Suppression of a Field Population of *Aedes aegypti* in Brazil by Sustained Release of Transgenic Male Mosquitoes," *PLoS Neglected Tropical Diseases* 9, no. 7 (July 2, 2015): e0003864, doi.org /10.1371/journal.pntd.0003864.

14. Heidi Ledford and Ewen Callaway, "Pioneers of Revolutionary CRISPR Gene Editing Win Chemistry Nobel," *Nature* 586, no. 7829 (October 7, 2020): 346–47, nature.com/articles/d41586-020-02765-9.

15. Shanthinie Ashokkumar et al., "Creation of Novel Alleles of Fragrance Gene OsBADH2 in Rice through CRISPR/Cas9 Mediated Gene Editing," *PLoS ONE* 15, no. 8 (August 12, 2020): e0237018, doi.org/10.1371 /journal.pone.0237018.

16. Andrew Scott, "How CRISPR Is Transforming Drug Discovery," *Nature* 555, no. 7695 (March 7, 2018): S10–11, nature.com/articles/d41586-018 -02477-1.

17. Sudarshan S. Lakhawat et al., "Implications of CRISPR-Cas9 in Developing Next Generation Biofuel: A Mini-Review," *Current Protein and Peptide Science* 23, no. 9 (2022): 274–584, pubmed.ncbi.nlm.nih.gov/36082852/.

18. Kevin V. Pixley et al., "Genome Editing, Gene Drives, and Synthetic Biology: Will They Contribute to Disease-Resistant Crops, and Who Will Benefit?," *Annual Review of Phytopathology* 57, no. 1 (2019): 165–88, doi.org /10.1146/annurev-phyto-080417-045954; Rubén Mateos Fernández et al., "Insect Pest Management in the Age of Synthetic Biology," *Plant Biotech-*

nology Journal 20, no. 1 (January 2022): 25–36, doi.org/10.1111/pbi.13685; Suzanne I. Warwick, Hugh J. Beckie, and Linda M. Hall, "Gene Flow, Invasiveness, and Ecological Impact of Genetically Modified Crops," *Annals of the New York Academy of Sciences* 1168, no. 1 (2009): 72–99, doi.org /10.1111/j.1749-6632.2009.04576.x.

19. Michael Kalos and Carl H. June, "Adoptive T Cell Transfer for Cancer Immunotherapy in the Era of Synthetic Biology," *Immunity* 39, no. 1 (July 25, 2013): doi.org/10.1016/j.immuni.2013.07.002.

20. Michael Eisenstein, "Gene Therapies Close in on a Cure for Sickle-Cell Disease," *Nature* 596 (August 25, 2021): S1–4, nature.com/articles/d41586 -021-02138-w.

21. Eisenstein, "Fix the Gene, Cure the Disease."

22. The Royal Swedish Academy of Sciences, "Press Release: The Nobel Prize in Chemistry 2020," NobelPrize.org, October 7, 2020, nobelprize.org /prizes/chemistry/2020/press-release/.

23. Scott Horsley, "How the U.S. Got into This Baby Formula Mess," NPR, May 19, 2022, sec. Business, npr.org/2022/05/19/1099748064/baby-infant -formula-shortages.

24. Emily Waltz, "First Genetically Modified Mosquitoes Released in the United States," *Nature* 593, no. 7858 (May 3, 2021): 175–76, doi.org/10.1038/d41586 -021-01186-6.

25. US Food and Drug Administration, "Oxitec Mosquito," FDA, August 5, 2017, fda.gov/animal-veterinary/intentional-genomic-alterations-igas-animals/ oxitec-mosquito; Waltz, "First Genetically Modified Mosquitoes Released in the United States."

26. US Environmental Protection Agency, "Following Review of Available Data and Public Comments, EPA Expands and Extends Testing of Genetically Engineered Mosquitoes to Reduce Mosquito Populations," Announcements and Schedules, March 7, 2022, epa.gov/pesticides/following-review-available -data-and-public-comments-epa-expands-and-extends-testing.

27. "Florida Keys," Oxitec, January 11, 2024, oxitec.com/florida-project.

28. Institute of Medicine, "The Background of Smoking Bans," in *Secondhand Smoke Exposure and Cardiovascular Effects: Making Sense of the Evidence* (Washington, DC: The National Academies Press, 2010), doi.org/10.17226 /12649.

29. Institute of Medicine, "The Background of Smoking Bans."

30. Mila de Mier, "Tell the EPA NO Genetically Modified Mosquitoes!," Change.org, April 3, 2012, change.org/p/tell-the-epa-no-to-gmo-mosquitoes.

31. Gillian Mohney and Justine Quart, "Fighting Zika in the US: The Battle Over GMO Mosquitoes," ABC News, July 5, 2016, https://abcnews.go .com/Health/deepdive/fighting-zika-40277607.

32. "NO to GM Mosquitoes in the Florida Keys," Facebook, accessed February 2, 2024, facebook.com/NoToGMMosquitoes.

33. Kacey C. Ernst et al., "Awareness and Support of Release of Genetically Modified 'Sterile' Mosquitoes, Key West, Florida, USA," *Emerging Infectious Diseases* 21, no. 2 (February 2015): 320–24, doi.org/10.3201/eid2102.141035.

34. Lizette Alvarez, "A Mosquito Solution (More Mosquitoes) Raises Heat in Florida Keys," *New York Times*, February 19, 2015, nytimes.com/2015/02 /20/us/battle-rises-in-florida-keys-over-fighting-mosquitoes-with -mosquitoes.html.

35. "City of Key West, FL—Meeting of City Commission on 4/3/2012 at 6:00 PM," accessed February 1, 2024, keywest.legistar.com/MeetingDetail .aspx?ID=192357&GUID=5E61409A-390E-4919-A7C0-39B7A7 EEF57E&Options=&Search=.

36. World Health Organization, "Zika Virus," December 8, 2022, who.int /news-room/fact-sheets/detail/zika-virus.

37. Duane J. Gubler et al., "History and Emergence of Zika Virus," *The Journal of Infectious Diseases* 216, no. Suppl 10 (December 15, 2017): S860–67, doi.org/10.1093/infdis/jix451; World Health Organization, "The History of Zika Virus," February 7, 2016, who.int/news-room/feature-stories/detail /the-history-of-zika-virus.

38. Gubler et al., "History and Emergence of Zika Virus."

39. Declan Butler, "What First Case of Sexually Transmitted Ebola Means for Public Health," *Nature*, October 16, 2015, doi.org/10.1038/nature.2015 .18584.

40. Frank Fenner, ed., *Smallpox and Its Eradication*, History of International Public Health, no. 6 (Geneva: World Health Organization, 1988).

41. Dimie Ogoina et al., "The 2017 Human Monkeypox Outbreak in Nigeria— Report of Outbreak Experience and Response in the Niger Delta University Teaching Hospital, Bayelsa State, Nigeria," *PLOS ONE* 14, no. 4 (April 17, 2019): e0214229, doi.org/10.1371/journal.pone.0214229.

42. Pan American Health Organization, "17 November 2015: Increase of Microcephaly in the Northeast of Brazil—Epidemiological Alert," November 17, 2015, paho.org/en/documents/17-november-2015-increase-microcephaly-northeast-brazil-epidemiological-alert-1.

43. Pan American Health Organization, "17 November 2015: Increase of Microcephaly in the Northeast of Brazil—Epidemiological Alert."

44. European Centre for Disease Control and Prevention, "Rapid Risk Assessment: Microcephaly in Brazil Potentially Linked to the Zika Virus Epidemic, 25 November 2015" (European Centre for Disease Control and Prevention, November 25, 2015), ecdc.europa.eu/en/publications-data/rapid-risk-assessment-microcephaly-brazil-potentially-linked-zika-virus-epidemic.

45. Centers for Disease Control and Prevention, "Pregnancy and Rubella," December 31, 2020, cdc.gov/rubella/pregnancy.

46. Michelle Morales et al., "Rubella," in *CDC Yellow Book: Health Information for International Travel: 2024* (National Center for Emerging and Zoonotic Infectious Diseases [NCEZID], Division of Global Migration Health [DGMH], 2023), wwwnc.cdc.gov/travel/yellowbook/2024/infections-diseases/rubella.

47. Laura A. Zimmerman, "Progress Toward Rubella and Congenital Rubella Syndrome Control and Elimination—Worldwide, 2012–2020," *Morbidity and Mortality Weekly Report* 71 (2022), doi.org/10.15585/mmwr.mm7106a2.

48. Centers for Disease Control and Prevention, "CMV Fact Sheet for Pregnant Women and Parents," n.d., stacks.cdc.gov/view/cdc/82562.

49. Centers for Disease Control and Prevention, "CMV Fact Sheet for Pregnant Women and Parents."

50. Megan H. Pesch et al., "Congenital Cytomegalovirus Infection," *BMJ: British Medical Journal* (Online) 373 (June 3, 2021), doi.org/10.1136/bmj.n1212.

51. Centers for Disease Control and Prevention, "Additional Area of Active Zika Transmission Identified in Miami Beach," August 19, 2016, archive.cdc.gov/#/details?url=https://www.cdc.gov/media/releases/2016/p0819-zika-miami-beach.html.

52. "Wynwood A 'Ghost Town' As Spraying For Zika Continues," CBS News Miami, August 7, 2016, cbsnews.com/miami/news/insecticide-rains-down-on-miamis-zika-zone/.

53. Kelly Servick, "Update: Florida Voters Split on Releasing GM Mosquitoes," *Science*, November 10, 2016, science.org/content/article/update-florida-voters -split-releasing-gm-mosquitoes.

54. Florida Keys Mosquito Control District, "FKMCD Board Approves Oxitec Mosquito Pilot Project in Florida Keys," August 19, 2020, keysmos quito.org/wp-content/uploads/2020/08/GMM-FKMCD-Board-FINAL-1 .pdf.

55. Alvarez, "A Mosquito Solution (More Mosquitoes) Raises Heat in Florida Keys."

56. Waltz, "First Genetically Modified Mosquitoes Released in the United States."

57. Oxitec, "Next Phase of Florida Keys Mosquito Control District and Oxitec Pilot Project Set to Commence in the Florida Keys," September 9, 2022, oxitec .com/en/news/next-phase-of-florida-keys-pilot-project-set-to-commence -in-the-florida-keys.

58. Oxitec, "Next Phase of Florida Keys Mosquito Control District and Oxitec Pilot Project Set to Commence in the Florida Keys."

59. Florida Keys Mosquito District, "MosquitoMate Wolbachia Trial," n.d., keysmosquito.org/mosquito-mate-wolbachia-trial/.

60. Cynthia E. Schairer et al., "Oxitec and MosquitoMate in the United States: Lessons for the Future of Gene Drive Mosquito Control," *Pathogens and Global Health* 115, no. 6 (August 18, 2021): 365–76, doi.org/10.1080/20477724 .2021.1919378.

61. Oxitec, "Oxitec Successfully Completes First Field Deployment of 2nd Generation Friendly™ Aedes Aegypti Technology," June 3, 2019, oxitec. com/en/news/oxitec-successfully-completes-first-field-deployment-of-2nd -generation-friendly-aedes-aegypti-technology; Taylor White, "First GMO Mosquitoes to Be Released in the Florida Keys," Genetic Engineering and Society (GES) Center, April 12, 2021, ges.research.ncsu.edu/ges/2021/04 /first-gmo-mosquitoes-to-be-released-in-the-florida-keys-undark/.

62. Richard Harris, "CDC Reveals Sharper Numbers of Zika Birth Defects From U.S. Territories," NPR, June 8, 2017, sec. Public Health, npr.org/sec tions/health-shots/2017/06/08/532087184/cdc-reveals-sharper-numbers -of-zika-birth-defects-from-u-s-territories; Helen Branswell, "Puerto Rico Declares Its Outbreak of Zika Virus Is Over," *STAT* (blog), June 5, 2017, statnews.com/2017/06/05/puerto-rico-zika-outbreak/; Helen Branswell, "Feud

Erupted between CDC, Puerto Rico over Reporting of Zika Cases, Document Shows," *STAT* (blog), May 1, 2017, statnews.com/2017/05/01/zika-virus -puerto-rico-cdc/.

63. Centers for Disease Control and Prevention, "Yellow Fever," September 14, 2018, archive.cdc.gov/#/details?url=https://www.cdc.gov/globalhealth /newsroom/topics/yellowfever/index.html.

64. World Health Organization, "Fractional Dose Yellow Fever Vaccine as a Dose-Sparing Option for Outbreak Response: WHO Secretariat Information Paper" (Geneva, Switzerland: World Health Organization, 2016), apps .who.int/iris/handle/10665/246236.

65. World Health Organization, "Fractional Dose Yellow Fever Vaccine as a Dose-Sparing Option for Outbreak Response."

66. UNICEF, "Malaria in Africa," UNICEF DATA, 2023, data.unicef.org /topic/child-health/malaria/.

67. US Government Accountability Office, "Overseas Nuclear Material Security: A Comprehensive National Strategy Could Help Address Risks of Theft and Sabotage," March 30, 2023, gao.gov/products/gao-23-106486.

68. International Atomic Energy Agency, "Nuclear Safety and Security," n.d., iaea.org/topics/nuclear-safety-and-security; US Nuclear Regulatory Commission, "Nuclear Security and Safeguards," NRC Web, October 6, 2021, nrc.gov/security.html.

69. Donald Mahley, "Opening Statement," Biological Weapons Convention (BWC) Annual Meeting of States Parties, Geneva, Switzerland, Department of State. The Office of Electronic Information, Bureau of Public Affairs, December 5, 2005, 2001-2009.state.gov/t/ac/rls/rm/58069.htm.

70. Yang Xue, "Developing Code of Conduct for Biology Scientists Under the Framework of Convention," BWC Meeting of Experts MX2, Geneva, Switzerland, November 2020, documents.unoda.org/wp-content/uploads /2020/11/CBRS-YANGXUE2020GenevaMX2.pdf; Submitted by China, "Proposal for the Development of the Template of Biological Scientist Code of Conduct under the Biological Weapons Convention," Member State Proposal BWC/MSP/2015/WP.9 (Geneva, Switzerland, December 15, 2015), unoda-documents-library.s3.amazonaws.com/Biological_Weapons_Con vention_-_Meeting_of_States_Parties_(2015)/Chinese%2B%28Unoffi cial%2BEnglish%2Btranslation%2Bannexed%29.pdf.

CHAPTER 10: MYSTERIES

1. Centers for Disease Control and Prevention, "QuickStats: Death Rates from Influenza and Pneumonia Among Persons Aged ≥65 Years, by Sex and Age Group—National Vital Statistics System, United States, 2018," *Morbidity and Mortality Weekly Report* 69 (2020), doi.org/10.15585/mmwr.mm6940a5.

2. Diya Surie, "Disease Severity of Respiratory Syncytial Virus Compared with COVID-19 and Influenza Among Hospitalized Adults Aged ≥60 Years—IVY Network, 20 U.S. States, February 2022—May 2023," *Morbidity and Mortality Weekly Report* 72 (2023), doi.org/10.15585/mmwr.mm7240a2.

3. Diane E. Pappas and J. Owen Hendley, "The Common Cold," *Principles and Practice of Pediatric Infectious Disease*, 2008, 203–6, doi.org/10.1016/B978-0-7020-3468-8.50034-1.

4. You Li et al., "Global Patterns in Monthly Activity of Influenza Virus, Respiratory Syncytial Virus, Parainfluenza Virus, and Metapneumovirus: A Systematic Analysis," *The Lancet Global Health* 7, no. 8 (August 1, 2019): e1031–45, doi.org/10.1016/S2214-109X(19)30264-5.

5. Lisa M. Casanova et al., "Effects of Air Temperature and Relative Humidity on Coronavirus Survival on Surfaces," *Applied and Environmental Microbiology* 76, no. 9 (May 2010): 2712–17, doi.org/10.1128/AEM.02291-09.

6. Rob Stein, "The CDC Sees Signs of a Late Summer COVID Wave," NPR, July 28, 2023, sec. Public Health, npr.org/sections/health-shots/2023/07/28/1190443473/the-cdc-sees-signs-of-a-late-summer-covid-wave; Jeffrey P. Townsend et al., "Seasonality of Endemic COVID-19," *mBio* 14, no. 6 (November 8, 2023): e01426-23, doi.org/10.1128/mbio.01426-23.

7. Ross Beall and Bill McNary, "Nearly 90% of U.S. Households Used Air Conditioning in 2020" (Washington, DC: U.S. Energy Information Administration, May 31, 2022), eia.gov/todayinenergy/detail.php?id=52558.

8. Stephen R. Kellert et al., "Adults Spend Little Time Outside Weekly," *The Nature of Americans* (Mishawaka, IN: DJ Case & Associates, April 2017), natureofamericans.org/findings/viz/adults-spend-little-time-outside-weekly.

9. Leslie A. Hoffman and Joel A. Vilensky, "Encephalitis Lethargica: 100 Years after the Epidemic," *Brain* 140, no. 8 (August 1, 2017): 2246–51, doi.org/10.1093/brain/awx177.

10. JMS Pearce, "Epidemic Encephalitis Lethargica," *Hektoen International*, April 29, 2020, hekint.org/2020/04/29/epidemic-encephalitis-lethargica/.

11. Laurie Winn Carlson, *A Fever in Salem: A New Interpretation of the New England Witch Trials* (Maryland: Rowman & Littlefield, 2000).

12. Ankit Vyas and Orlando De Jesus, "Von Economo Encephalitis," in *Stat-Pearls* (Treasure Island, FL: StatPearls Publishing, 2024), ncbi.nlm.nih.gov/books/NBK567791/.

13. Hoffman and Vilensky, "Encephalitis Lethargica."

14. Hoffman and Vilensky, "Encephalitis Lethargica."

15. Hoffman and Vilensky, "Encephalitis Lethargica."

16. National Institute of Neurological Disorders and Stroke, "Post-Polio Syndrome," November 28, 2023, ninds.nih.gov/health-information/disorders/post-polio-syndrome.

17. Centers for Disease Control and Prevention, "Shingles: Cause and Transmission," May 10, 2023, cdc.gov/shingles/about/.

18. Kjetil Bjornevik et al., "Epstein–Barr Virus as a Leading Cause of Multiple Sclerosis: Mechanisms and Implications," *Nature Reviews Neurology* 19, no. 3 (March 2023): 160–71, doi.org/10.1038/s41582-023-00775-5; National Institutes of Health (NIH), "Study Suggests Epstein-Barr Virus May Cause Multiple Sclerosis," *NIH Research Matters*, February 1, 2022, nih.gov/news-events/nih-research-matters/study-suggests-epstein-barr-virus-may-cause-multiple-sclerosis.

19. Oliver Sacks, *Awakenings* (London: Duckworth & Co., 1973).

20. Russell C. Dale et al., "Encephalitis Lethargica Syndrome: 20 New Cases and Evidence of Basal Ganglia Autoimmunity," *Brain* 127, no. 1 (January 1, 2004): 21–33, doi.org/10.1093/brain/awh008.

21. Sherman McCall et al., "The Relationship between Encephalitis Lethargica and Influenza: A Critical Analysis," *Journal of Neurovirology* 14, no. 3 (May 2008): 177–85, doi.org/10.1080/13550280801995445.

22. Joel A. Vilensky et al., "A Historical Analysis of the Relationship between Encephalitis Lethargica and Postencephalitic Parkinsonism: A Complex Rather than a Direct Relationship," *Movement Disorders* 25, no. 9 (2010): 1116–23, doi.org/10.1002/mds.22908.

23. Jennie Johnstone et al., "Viral Infection in Adults Hospitalized with Community-Acquired Pneumonia: Prevalence, Pathogens, and Presentation,"

Chest 134, no. 6 (December 1, 2008): 1141–48, doi.org/10.1378/chest
.08-0888.

24. Center for Cancer Research, National Cancer Institute, "New Tool Cata-
logs Thousands of Previously Unknown Viruses," February 4, 2020, ccr
.cancer.gov/news/article/new-tool-catalogs-thousands-of-previously
-unknown-viruses; Jim Robbins, "Before the Next Pandemic, an Ambitious
Push to Catalog Viruses in Wildlife," Yale Environment 360, April 22,
2020, e360.yale.edu/features/before-the-next-pandemic-an-ambitious-push
-to-catalog-viruses-in-wildlife; Simon J. Anthony et al., "A Strategy to Estimate Unknown Viral Diversity in Mammals," *mBio* 4, no. 5 (September
3, 2013): doi.org/10.1128/mbio.00598-13.

25. "Development Data Library," USAID Development Data Library (DDL),
accessed February 2, 2024, data.usaid.gov/browse?q=predict.

26. Katendi Changula et al., "Serological Evidence of Filovirus Infection in
Nonhuman Primates in Zambia," *Viruses* 13, no. 7 (July 2021): 1283, doi
.org/10.3390/v13071283.

27. Dimitrios Gouglas et al., "The 100 Days Mission—2022 Global Pandemic
Preparedness Summit," *Emerging Infectious Diseases* 29, no. 3 (March 2023):
e221142, doi.org/10.3201/eid2903.221142; Melanie Saville et al., "Delivering
Pandemic Vaccines in 100 Days—What Will It Take?," *New England Journal
of Medicine* 387, no. 2 (July 14, 2022): e3, doi.org/10.1056/NEJMp2202669.

28. Dimitrios Gouglas et al., "Estimating the Cost of Vaccine Development
against Epidemic Infectious Diseases: A Cost Minimisation Study," *The
Lancet Global Health* 6, no. 12 (December 1, 2018): e1386–96, doi.org
/10.1016/S2214-109X(18)30346-2.

29. David B. Fogel, "Factors Associated with Clinical Trials That Fail and Opportunities for Improving the Likelihood of Success: A Review," *Contemporary Clinical Trials Communications* 11 (August 7, 2018): 156–64, doi
.org/10.1016/j.conctc.2018.08.001.

30. World Health Organization, "Ebola Virus Disease," 2021, who.int/news
-room/fact-sheets/detail/ebola-virus-disease.

31. "The World of Air Transport in 2019," accessed January 29, 2024, icao
.int/annual-report-2019/Pages/the-world-of-air-transport-in-2019.aspx.

32. Dweepobotee Brahma et al., "The Early Days of a Global Pandemic: A
Timeline of COVID-19 Spread and Government Interventions," Brookings,

April 2, 2020, brookings.edu/articles/the-early-days-of-a-global-pandemic
-a-timeline-of-covid-19-spread-and-government-interventions/.

33. Donald Rumsfeld, *Known and Unknown: A Memoir* (New York: Sentinel, 2011).

CONCLUSION: PROGRESS

1. Alan Feuer and Andrea Salcedo, "New York City Deploys 45 Mobile Morgues as Virus Strains Funeral Homes," *New York Times*, April 10, 2020, nytimes .com/2020/04/02/nyregion/coronavirus-new-york-bodies.html.

2. Sharon Otterman, "Why 530 Frozen Bodies Sit in a Brooklyn Warehouse," *New York Times*, December 26, 2020, sec. New York, nytimes.com/2020 /12/26/nyregion/frozen-bodies-coronavirus-brooklyn.html.

3. Michael Schwirtz, "USNS Comfort Hospital Ship Was Supposed to Aid New York. It Has 20 Patients.," *New York Times*, April 2, 2020, nytimes .com/2020/04/02/nyregion/ny-coronavirus-usns-comfort.html.

4. Todd C. Lopez, "Comfort, Javits Center Open Care to COVID-19 Patients," Department of Defense News, April 7, 2020, https://www.defense .gov/News/News-Stories/article/article/2140535/comfort-javits-center -open-care-to-covid-19-patients/.

5. Helen Branswell, "Why 'Flattening the Curve' May Be the World's Best Bet to Slow the Coronavirus," *STAT* (blog), March 11, 2020, statnews .com/2020/03/11/flattening-curve-coronavirus/.

6. Stephanie Kramer, "More Americans Say They Are Regularly Wearing Masks in Stores and Other Businesses," Pew Research Center, August 27, 2020, pewresearch.org/short-reads/2020/08/27/more-americans-say-they-are -regularly-wearing-masks-in-stores-and-other-businesses/.

7. Andrew Daniller, "Americans Remain Concerned That States Will Lift Restrictions too Quickly, but Partisan Differences Widen," Pew Research Center, May 7, 2020, pewresearch.org/short-reads/2020/05/07/americans -remain-concerned-that-states-will-lift-restrictions-too-quickly-but-partisan -differences-widen/.

8. Imen Ayouni et al., "Effective Public Health Measures to Mitigate the Spread of COVID-19: A Systematic Review," *BMC Public Health* 21 (May 29, 2021): 1015, doi.org/10.1186/s12889-021-11111-1.

9. Lawrence O. Gostin, "Fix the Backlash against Public Health," *Science* 379, no. 6639 (March 30, 2023): 1277.

10. "Underfunded and Under Threat Archives," *KFF Health News* (blog), accessed February 2, 2024, kffhealthnews.org/news/tag/underfunded-and -under-threat/.

11. Cary Funk et al., "Americans' Largely Positive Views of Childhood Vaccines Hold Steady," Pew Research Center, May 16, 2023, pewresearch.org/sci ence/2023/05/16/americans-largely-positive-views-of-childhood-vaccines -hold-steady/.

12. Tim Lancaster and Lindsay F. Stead, "Individual Behavioural Counselling for Smoking Cessation," *Cochrane Database of Systematic Reviews*, no. 3 (2017), doi.org/10.1002/14651858.CD001292.pub3.

13. Bao Yen Luong Thanh et al., "Behavioural Interventions to Promote Workers' Use of Respiratory Protective Equipment," *Cochrane Database of Systematic Reviews* 12 (2016), doi.org/10.1002/14651858.CD010157.pub2.

14. Joanne L. Jordan et al., "Interventions to Improve Adherence to Exercise for Chronic Musculoskeletal Pain in Adults," *Cochrane Database of Systematic Reviews*, no. 1 (2010), doi.org/10.1002/14651858.CD005956.pub2.

15. Nipun Shrestha et al., "Workplace Interventions for Reducing Sitting at Work," *Cochrane Database of Systematic Reviews*, no. 12 (2018), doi.org/10 .1002/14651858.CD010912.pub5.

16. Eileen F. S. Kaner et al., "Effectiveness of Brief Alcohol Interventions in Primary Care Populations," *Cochrane Database of Systematic Reviews*, no. 2 (2018), doi.org/10.1002/14651858.CD004148.pub4.

17. Sze Lin Yoong et al., "Healthy Eating Interventions Delivered in Early Childhood Education and Care Settings for Improving the Diet of Children Aged Six Months to Six Years," *Cochrane Database of Systematic Reviews*, no. 6 (2023), https://doi.org/10.1002/14651858.CD013862.pub2.

18. "Breast Milk Is Best," Johns Hopkins Medicine, December 8, 2021, hop kinsmedicine.org/health/conditions-and-diseases/breastfeeding-your -baby/breast-milk-is-the-best-milk.

19. Allan Walker, "Breast Milk as the Gold Standard for Protective Nutrients," *The Journal of Pediatrics* 156, no. 2 (February 1, 2010): S3–7, doi.org/10 .1016/j.jpeds.2009.11.021.

20. Mary Bresnahan et al., "Made to Feel Like Less of a Woman: The Experi-

ence of Stigma for Mothers Who Do Not Breastfeed," *Breastfeeding Medicine* 15, no. 1 (January 2020): 35–40, doi.org/10.1089/bfm.2019.0171.

21. Centers for Disease Control and Prevention, "CDC's Infant and Toddler Nutrition Website," June 3, 2022, cdc.gov/nutrition/infantandtoddlernu trition/index.html.

22. J. W. McKenna and K. N. Williams, "Crafting Effective Tobacco Counteradvertisements: Lessons from a Failed Campaign Directed at Teenagers," *Public Health Reports* 108, no. Suppl 1 (1993): 85–89.

23. Brittaney Kiefer, "Cancer Research UK Links Obesity to Cancer with Ads Resembling Cigarette Packs," Campaign US, July 2, 2019, campaignlive .com/article/cancer-research-uk-links-obesity-cancer-ads-resembling -cigarette-packs/1589767.

24. Prabhat Jha et al., "21st-Century Hazards of Smoking and Benefits of Cessation in the United States," *New England Journal of Medicine* 368, no. 4 (January 24, 2013): 341–50, doi.org/10.1056/NEJMsa1211128.

25. Mike Stobbe, "U.S. Adult Cigarette Smoking Hits New All-Time Low," *PBS NewsHour*, April 27, 2023, pbs.org/newshour/health/u-s-adult-cigarette -smoking-hits-new-all-time-low.

26. Lisa Bowleg, "We're Not All in This Together: On COVID-19, Intersectionality, and Structural Inequality," *AJPH* 110, no. 7 (July 2020), ajph .aphapublications.org/doi/full/10.2105/AJPH.2020.305766.

27. World Health Organization, "Ebola Outbreak 2022—Uganda," 2022, who .int/emergencies/situations/ebola-uganda-2022.

28. World Health Organization, "Cholera—Global Situation," Disease Outbreak News, February 11, 2023, who.int/emergencies/disease-outbreak-news/item /2023-DON437.

29. Ed Yong, "We Created the 'Pandemicene,'" *Atlantic*, April 28, 2022, theatlantic.com/science/archive/2022/04/how-climate-change-impacts -pandemics/629699/.

INDEX